国匠承启卷——
传统民居保护性利用设计

Craftsmanship Heritage —
Design for the Protective Utilization of Traditional Folk Houses

中国建筑学会室内设计分会
北京建筑大学 编

CIID室内设计教学参考书 CIID"室内设计6+1"2017(第五届)校企联合毕业设计

出版支持：北京建筑大学教材建设专项

中国水利水电出版社
www.waterpub.com.cn
·北京·

内 容 提 要

做好地域代表性民居保护和利用，培育中国工匠，大力弘扬传统建筑工匠精神，厚植传统建筑工匠文化，加强传统建造技术的传承发展，是作为贯彻党中央、国务院《关于实施中华优秀传统文化传承工程的意见》的重要举措。为在室内设计师教育培养中探讨"传统民居保护性利用设计"问题，由中国建筑学会室内设计分会主办了CIID"室内设计6+1"2017（第五届）校企联合毕业设计活动。

"室内设计6+1"校企联合毕业设计是中国建筑学会室内设计分会于2013年创立的设计教育示范项目。本届活动由同济大学、华南理工大学、哈尔滨工业大学、西安建筑科技大学、北京建筑大学、南京艺术学院、浙江工业大学等作为参加高校并分别联合企业共同指导毕业设计。来自建筑学、环境设计、艺术与科技等专业的2017届毕业班师生们，通过学会、高校、企业、专家间的协同，就"传统民居保护性利用设计"总命题下的民居保护规划与景观设计、民居建筑保护与更新设计、营造技艺传承与展陈设计等课题方向，联合开展了综合性实践教学活动。师生们对民居保护、开发利用、有机更新、技艺传承等问题的探讨感同身受，交流深入，促进了毕业设计教学水平和人才培养质量的提升。据此，中国建筑学会室内设计分会和北京建筑大学联合编写了中国建筑学会室内设计分会推荐专业教学参考书《国匠承启卷——传统民居保护性利用设计》。全书包括活动规章、调研踏勘、中期检查、答辩展示、教学研究、专家讲坛、历史定格等章节，采用中英文对照方式，记载了本届设计教育项目的开展情况和成果，图文并茂、内容翔实。

本书可供建筑学、环境设计、艺术与科技、城乡规划、风景园林、城市地下空间工程、工业设计、产品设计、视觉传达设计以及公共艺术等专业人员及设置相关专业的院校师生参考借鉴。

图书在版编目（CIP）数据

国匠承启卷：传统民居保护性利用设计 / 中国建筑学会室内设计分会，北京建筑大学编. -- 北京：中国水利水电出版社，2017.10
CIID室内设计教学参考书：CIID"室内设计6+1"2017（第五届）校企联合毕业设计
ISBN 978-7-5170-5998-1

Ⅰ. ①国… Ⅱ. ①中… ②北… Ⅲ. ①住宅－室内装饰设计－毕业设计－高等学校－教学参考资料 Ⅳ. ①TU241

中国版本图书馆CIP数据核字(2017)第262306号

书　名	CIID室内设计教学参考书 CIID"室内设计6+1"2017（第五届）校企联合毕业设计 **国匠承启卷———传统民居保护性利用设计** GUOJIANG CHENGQI JUAN---CHUANTONG MINJU BAOHUXING LIYONG SHEJI
作　者	中国建筑学会室内设计分会　北京建筑大学　编
出版发行	中国水利水电出版社 （北京市海淀区玉渊潭南路1号D座　100038） 网址：www.waterpub.com.cn E-mail：sales@waterpub.com.cn 电话：（010）68367658（营销中心）
经　售	北京科水图书销售中心（零售） 电话：（010）88383994、63202643、68545874 全国各地新华书店和相关出版物销售网点
排　版	中国建筑学会室内设计分会
印　刷	北京博图彩色印刷有限公司
规　格	210mm×285 mm　16开本　16印张　855千字
版　次	2017年10月第1版　2017年10月第1次印刷
印　数	0001—1000册
定　价	120.00元

凡购买我社图书，如有缺页、倒页、脱页的，本社营销中心负责调换
版权所有·侵权必究

编委会
Editorial Committee

Director   主任
Zou Huying, Li Aiqun, Zhang Dayu   邹瑚莹 李爱群 张大玉

Deputy Director   副主任
Ye Hong, Su Dan, Li Junqi, Zou Jiting   叶 红 苏 丹 李俊奇 邹积亭
Li Zhenyu, Zhang Ke, Mei Hongyuan, Lin Baogang   李振宇 张 珂 梅洪元 蔺宝钢
Hu Xuesong, Li Yiwen, Wang Jiansheng   胡雪松 李亦文 王建胜

Committee member   委员
Zhao Jian, Wang Weimin, Huang Quan, Yao Ling   赵 健 王炜民 黄 全 姚 领
Zhu Haixuan, Kou Jianchao, Cui Lin, Zeng Jun   朱海玄 寇建超 崔 林 曾 军
Zhuo Pei, Chen Weixin, Wang Chuanshun, Guo Xiaoming   卓 培 陈卫新 王传顺 郭晓明
Zuo Yan, Xie Guanyi, Xue Ying, Luo Wen   左 琰 谢冠一 薛 颖 骆 雯
Shi Tuo, Zhou Lijun, Ma hui, Zhao Hui   石 拓 周立军 马 辉 兆 翚
Liu Xiaojun, Gu Qiulin, Yang Lin, Zhu Ningke   刘晓军 谷秋琳 杨 琳 朱宁克
Zhu Fei, Lv Qinzhi, Ren Yi, Song Yang   朱 飞 吕勤智 任 彝 宋 杨

Chief Editor   主编
Chen Jingyong   陈静勇

Associate editor   副主编
Yang Lin, Sun Xiaopeng, Pan Xiaowei   杨 琳 孙小鹏 潘晓微

Editorial Committee   执行编委
Zhu Ningke, Lin Zequan, He Nan   朱宁克 林则全 和 楠

Cover Design   封面设计
Lin Zequan   林则全

Graphic Design   装帧设计
Sun Xiaopeng, He Nan   孙小鹏 和 楠

Preface

Since its opening in 2013, the annual CIID "Interior Design 6+1" University-Enterprise Cooperative Graduation Design Program has been held for five years. Sponsored by CIID, created and joined by colleges and well-known design firms, the Program has been unfolded based on actual design problems to be resolved by the design firms. The design themes adopted over the years are Post-Game Business Opportunity—Interior Design of Post-game Transformation of National Stadium, A New Look of Intercity Rail—Environmental Design of Shanghai Metro Reformation, Legacy of Weaponry Factory—Environmental Design of Nanjing Chenguang 1865 Creative Park, Elderly Care Album—Environmental Transformation Design of Beijing Yaoyang International Elderly Apartment, and Lineage of Craftsmanship Album—Design for the Protective Utilization of Traditional Folk Houses respectively. From proposal survey, mid-term examination and graduation oral defense, students from different areas and teaching systems have broadened their horizon, gained a lot and achieved great results through exchanges with mentors and experts from different colleges. This society and market oriented program has given the students an opportunity to apply the theories they learn to reality before graduation, and prepared them for the role shift to adapt to the rapid social development.

One of the features of the Program is to unite colleges and well-known design firms to build an educational platform for the promotion of educational development of interior design. Every year the teachers and students from 7 colleges and representatives from famous design firms participate in the Program. This year a new exchange pattern featuring 7 colleges, 7 cooperative enterprises and 7 design projects on the basis of one college working with one enterprise in regard to the region's characteristic folk houses has been introduced. The expert judges (enterprise mentors) are involved in every aspect of the exchange to master the evaluation standards and graduation design progress. The teachers and students from the 7 colleges can thus exchange with and learn from each other and improve through this platform.

Over the past five years, delightful achievements that support the training of innovative and entrepreneurial talents and advancement of educational reform have been made. As the only academic group in the Chinese interior design circles, CIID has been committed to boosting the development of China's interior design, focusing on the growth of young people, and fulfilling its due responsibility by working pragmatically and unremittingly. In the future, the Program will continue to work with the industry (enterprises) to launch graduation design events and reflect the principle of education at the service of talent demands. Lessons will be drawn from the past and explorations made to improve the Program and accelerate the reform and improvement of interior design education.

Zou Huying
July, 2017

序

自 2013 年以来,由中国建筑学会室内设计分会(CIID)主办、高校与知名设计企业共同创建并参与的 CIID "室内设计 6+1" 校企联合毕业设计活动截止到今天已经成功举办五届了。每一届设计题目的选定都来自知名设计企业当前需要解决的实际课题,分别以"赛后商机——国家体育场赛后改造室内设计""城轨新境——上海地铁改造环境设计""兵工遗产——南京晨光 1865 创意产业园环境设计""鹤发医养卷——北京曜阳国际老年公寓环境改造设计""国匠承启卷——传统民居保护性利用设计"为题开展相关活动。从开题的现场勘踏,到中期检查,再到毕业答辩,来自不同地域、不同教学体系的同学们通过与不同院校评委老师、专家们的交流,视角和思路都得到了拓展,受益良多,取得了良好成绩。活动既面向社会,面向市场,又让学生在毕业之前能理论联系实际地掌握知识,为今后尽快转变角色进行工作打下基础,以适应当前社会的高速发展。

"室内设计 6+1"校企联合毕业设计的特色之一是打造了一个高校与知名设计企业共同创建的教育平台,这是一个推动室内设计教育发展的平台,每年由 7 所高校师生和知名设计企业组成并参与活动。今年,联合毕设课题是采取了 7 所高校分别联合企业,从所在区域特色民居中选定的,形成了"7 所参与高校＋7 家合作企业＋7 个类型课题"的新的交流格局。交流活动中评审专家(企业导师)全程参与,有利于全面把握评审的标准和毕设过程情况,使得 7 所高校师生在这个平台上互相交流、互相学习、共同提高。

五年来,"室内设计 6+1"校企联合毕业设计活动取得了良好的成果,这有助于创新、创业型人才的培养,有助于学校教育改革的推进。室内设计分会作为中国室内设计界唯一的学术团体,一直致力于推动我国室内设计事业的发展,重视青年学生的成长,不为虚名浮华,踏踏实实坚持不懈地努力工作,是我们应尽的责任。今后,"6+1"的活动将继续坚持依托行(企)业开展联合毕业设计教学,体现教育服务于人才培养需求的原则,还要总结经验,不断摸索,把这项活动更好地开办下去,以促进室内设计教育的改革与提高。

邹瑚莹
2017 年 7 月

清华大学　教授
Zou Huying
President of CIID
Professor of Tsinghua University

Foreword

Volume 5 Craftsmanship Heritage is to be published. This volume records the special discussion about the "Design for the Protective Utilization of Traditional Folk Houses" in this activity and commemorates the five-year history of the CIID "Interior Design 6+1"University-Enterprise Cooperative Graduation Design Program.

In 2013, under the guidance of the strategy of "build an innovative nation and strengthen the country with talents", facing the new target of cultivating interior design talents and based on China's new professional environment and the new demands for service industry development, CIID initiated the interior design education demonstration project - CIID "Interior Design 6+1"University-enterprise Cooperation Graduation Project with the support of six universities located in different regions and offering architecture, design science or other interior design courses and a famous design enterprise, including Tongji University (TJU), South China University of Technology (SCUT), Harbin Institute of Technology (HIT), Xi'an University of Architecture and Technology (XAUAT), Beijing University of Civil Engineering and Architecture (BUCEA) and Nanjing University of the Arts (NUA). The project is aimed at exploring the road of interior designer education and cultivation towards the "Excellence Initiative" target.

It's the platform of five sessions. There are "6+1" regional universities, "6+1" university-enterprise cooperation, "6+1"activity links, "6+1" columns, "6+1"... Five sessions of the CIID "Interior Design 6+1"University-enterprise Cooperation Graduation Project Activity have made "6+1" become a brand platform of this CIID's interior design education demonstration project and presented continuously enriched new connotation and constantly enhanced influence.

A history of five years. The design topics have been focusing on the hot issues the industry and the society are concerned about. The topic of the first session in 2013 is "Business Opportunities after the Game —Interior Design for the National Stadium After-the-Game Transformation", involving the problems about operation and utilization of large-scale stadiums after large-scale sports events. The topic of the second session in 2014 is "New Environment of Urban Rail – Environment Design for Shanghai Metro Transformation", discussing the problems about environment design for urban renewal and metro stations. The topic of the third session in 2015 is "Heritage of Weaponry Industry —Environment Design for Nanjing Chenguang 1865 Creative Industry Park", discussing the problems about protection and utilization of modern industrial heritage of famous historic and cultural cities. The topic of the fourth session in 2016 is "Medical Caring of the Aged – Environmental Construction Design for Beijing Yaoyang International Apartment for the Elderly", exploring the problem about how the existing medical care architectural space environment adapts to the new demands of the aging society for livability. The topic of the fifth session in 2017 is "Craftsmanship Heritage - Design for the Protective Utilization of Traditional Folk Houses", comparing the problems about the protection and utilization of representative folk houses in multiple regions.

As the fifth session, this activity continues to insist on joining hands with enterprises to carry out cooperative graduation project teaching and reflecting the principle of serving the industry needs for talent cultivation. Through universities' independent contact and filing with CIID, this session tries the new mode of "777"under the guidance of the general proposition "Framework Design Brief", i.e. seven universities, seven enterprises and seven featured folk houses, highlighting the new positioning of "general proposition, unique features, joint guide, and service needs".

In 2017, to implement the guiding principle of the Opinions on Carrying out the Traditional Chinese Cultural Inheritance Projects of the Party Central Committee and the State Council - cultivating a number of representative personages of Chinese culture, implement the requirement of the Government Work Report 2017 on cultivating Chinese craftsmen, vigorously promote traditional architectural craftsmanship, inherit traditional architectural craftsman culture and strengthen inheritance and development of traditional construction technology, the Village and Town Construction Department of the Ministry of Housing and Urban-Rural Development organized to conduct the Chinese traditional architectural craftsman identification work.

Traditional architectural craftsmen, as the inheritors and practitioners of traditional construction technology, shoulder the historical mission of inheriting and continuing traditional construction techniques and renovation techniques. In the recent decades, as modern construction technology has been rapidly developing, demands for traditional construction techniques have been withering, and traditional construction craftsman teams have disappeared quickly. Traditional building renovation is facing adverse conditions, and many traditional construction techniques have failed to be handed down from past generations. Thus, carrying out traditional architectural craftsman identification work to excavate a group of excellent inheritors and representative personages of Chinese traditional construction technology is of great significance to enhancing the position of traditional construction craftsmen, revitalizing traditional construction craftsman teams, carrying forward traditional construction technology, comprehensively raising the level of traditional construction technology, promoting traditional architectural culture and strengthening national self-confidence and sense of pride.

According to the three topic directions -folk house protection planning and landscape design, folk house protection and update design, and construction technology inheritance and display design under the general proposition of"Design for the

前言

第5卷《国匠承启卷》即将付梓出版，本卷既是对本届活动"传统民居保护性利用设计"专题探讨的记录，更是CIID"室内设计6+1"校企联合毕业设计活动走过5年历程的纪念。

在建设创新型国家和人才强国战略的指引下，立足我国学科专业的新环境，面对建筑室内设计人才培养的新目标，服务行业发展的新需求，2013年，由中国建筑学会室内设计分会（CIID）主办，由同济大学、华南理工大学、哈尔滨工业大学、西安建筑科技大学、北京建筑大学、南京艺术学院等6所地处不同区域、设置建筑学或设计学等学科室内设计方向的高校与知名设计企业，共同创建了学会室内设计教育示范项目CIID"室内设计6+1"校企联合毕业设计，协同探索"卓越计划"目标下的室内设计师教育培养之路。

陈静勇
执行主编
中国建筑学会室内设计分会副理事长
北京建筑大学 教授

Chen JingYong
Executive Editor
Vice President of China Architecture
Society Interior Design Branch
Professor of Beijing Architecture University

5届的平台。有6+1所区域高校、6+1个校企联合、6+1个活动环节、6+1个编制栏目、6+1……，连续5届CIID"室内设计6+1"校企联合毕业设计活动使"6+1"这个室内设计教育示范项目成为品牌和平台，彰显出不断丰富的新内涵和不断提升的影响力。

5年的历程。设计选题始终是以行业、社会密切关注的热点问题为导向。2013（首届）的"赛后商机——国家体育场赛后改造室内设计"课题，涉及大型体育赛事之后大型体育场馆经营利用问题；2014（第二届）的"城轨新境——上海地铁改造环境设计"课题，探讨了城市更新与地铁站点的环境设计问题；2015（第三届）的"兵工遗产——南京晨光1865创意产业园环境设计"研究了历史文化名城近代工业遗产保护与利用问题；2016（第四届）的"鹤发医养——北京曜阳国际老年公寓环境改造设计"，挖掘了既有医养建筑空间环境适应老龄社会宜居新需求问题；2017（第五届）的"国匠承启——传统民居保护性利用设计"，比较了多地域代表性民居保护和利用的问题。

第5届活动。继续坚持联合企业开展联合毕业设计教学，体现服务于人才培养行业需求的原则。经参加高校自主联系并向学会备案，本届尝试了在总命题《框架任务书》指导下的7所高校、7家企业、7处特色民居的"777"式联合毕业设计活动的新方式，凸显活动"总体命题，纷显特色，联合指导，服务需求"的新定位。

2017年，为贯彻党中央、国务院《关于实施中华优秀传统文化传承工程的意见》提出的培养造就一批中华文化代表人物的精神，落实2017年《政府工作报告》关于培育中国工匠的要求，大力弘扬传统建筑工匠精神，厚植传统建筑工匠文化，加强传统建造技术的传承发展，由住房和城乡建设部村镇建设司组织开展了中国传统建筑名匠认定工作。

传统建筑工匠作为传统建造技术的传承者和实践者，担负着建筑建造、修缮和传统建造技术传承延续的历史使命。近几十年来，由于现代建筑技术的快速发展，传统建造技术需求持续萎缩，传统建筑工匠队伍消失迅速，大量的传统建筑修复面临困境，传统建造技术失传严重。开展中国传统建筑名匠认定工作，挖掘一批中国传统建造技术的优秀传承者和代表性人物，树杆立旗，轨物范世，对提升传统建筑工匠地位、复兴传统建筑工匠队伍、传承发扬传统建造技术、全面提升传统建造技术水平、弘扬传统建筑文化、增强民族自信心和自豪感具有重要意义。

通过学会、高校、企业、专家等多方协同，就"传统民居保护性利用设计"总命题下的民居保护规划与景观设计、民居建筑保护与更新设计、营造技艺传承与展陈设计3个课题方向，联合开展了综合性实践教学活动。师生们对民居保护、开发利用、有机更新、技艺传承等问题的探讨感同身受，交流深入，促进了毕业设计教学水平和人才培养质量的提升。

本届活动的6个主体环节是，2016年10月29日（杭州市，学会第26届年会）召开CIID"6+1室内设计"2017（第五届）校企联合毕业设计命题研讨会；2017年3月4日—5日（广州市，华南理工大学）开题仪式和广东民居考察；4月15日—16日（杭州市，浙江工业大学）中期检查和浙江民居考察；6月17日—

Protective comprehensive practical teaching activities. Teachers and students have had in-depth discussion about the problems with folk house protection, development and utilization, organic update and technique inheritance, which has facilitated the improvement of graduation project teaching level and the quality of talent cultivation.

This activity includes six main sections. On October 29, 2016 (at the 26th annual meeting of CIID, Hangzhou), CIID "6+1 Interior Design"- The 5th University-enterprise Cooperation Graduation Project Seminar 2017 was held; During March 4-5, 2017 (at SCUT, Guangzhou), proposal ceremony and Guangdong folk house investigation were conducted; During April 15-16, 2017 (at ZJUT, Hangzhou), medium-term inspection and Zhejiang folk house investigation were conducted; During June 17-18, 2017 (at BUCEA, Beijing) Beijing folk house investigation, defense review meeting and reward ceremony came to an end; CIID and BUCEA are responsible for collecting the works of the participating universities, enterprises, and lecture and review experts and editing and publishing the volume Craftsmanship Heritage- Design for the Protective Utilization of Traditional Folk Houses (Professional teaching reference book recommended by CIID: CIID "Interior Design 6+1"University-enterprise Cooperation Graduation Project; During November 8-10, 2017, exhibition on special topics will be presented at the 27th CIID (Jiangxi) Annual Meeting & International Academic Exchange Conference.

Craftsmanship Heritage, in both Chinese and English, was published at the same time as the 27th CIID (Jiangxi) Annual Meeting. This publication about the activity is one of the featured cases of CIID, AIDIA, IFI and other interior design international organizations in carrying out interior design international exchange and an extended link of this activity.

It's the concentration of five volumes. We have gradually developed the activity portfolio into teaching reference books recommended by CIID and formed the compilation culture and binding features of this series. The "6+1" columns in each volume cover the core contents of corresponding activity links, among which six columns such as Investigation and Survey, Medium-term Inspection, Defense Presentation, Teaching Research, Expert Rostrum, and History Freeze-frame record the main process and contents of this session, supplemented with expert comments, words from students, certificates of awards, photos of teachers and students etc. The Constitution (Revised in 2015) formed through five sessions' improvement and optimization, the Outline, the Graduation Project Framework Design Brief, and the Rules on the Defense, Appraisal and Commendation Work of this session have been collected in the first column Activity Rules, which is a long-term mechanism guiding the activity development and reflects that CIID's interior design education work has rules to follow.

In brief, I hope more activity contents can be recorded. Due to space limitation, careless omission may occur in edition. You are welcome to point out our mistakes so that we can make improvement.

Answering to the national strategy of innovation-driven development, serving the demand of socio-economic development and cultivating a group of high-quality interior designers who have strong innovation ability and meet the needs of the society is always a proposition that the society faces.

The 5th CIID "Interior Design 6+1"University-enterprise Cooperation Graduation Project Activity 2017 has ended in a satisfactory way. The graduates majoring in Architecture, Industrial Design, Environment Design, Arts, Science and Technology etc who have participated in the activity have been awarded Bachelor of Architecture, Bachelor of Engineering, or Bachelor of Arts. Some of them have become master degree candidates, some have gone overseas to pursue advanced studies, and some have got a job. Facing the industrial needs, they have embarked on a new journey of excellent development.

This year is the 28th anniversary of the establishment of CIID. Here, we would like to present Craftsmanship Heritage as the commemorative edition of the 5th anniversary of CIID's interior design education demonstration project- "Interior Design 6+1"University-enterprise Cooperation Graduation Project Activity to all friends. We hope all local specialized committees of CIID can join hands to further organize regional characteristic "Interior Design 6+1"university-enterprise Cooperation Graduation Project Activity and further expand brand exchange platforms so as to improve the quality of China's interior design talent cultivation.

We would like to thank the Professional Steering Committee for Architecture Discipline of Institutions of Higher Education in China, MOE Professional Teaching Committee for Design Science of Institutions of Higher Education in China and other organizations for their guidance on relevant discipline construction work of colleges and universities in China!

We also would like to acknowledge AAUA, the Organizing Committee of Asian Design Annual Award, all local professional committees of CIID, relevant colleges and universities, enterprises, lecture and review experts and publishing enterprises for supporting the CIID "Interior Design 6+1"University-enterprise Cooperation Graduation Project Activity!

At last, we would like to say thanks to all colleges and universities participating in the undertaking and construction of this activity and all volunteers devoted to this activity in the past five years!

The 2017/2018 autumn semester has began, and the freshmen of 2017 have entered school. Their works will be published on the commemorative edition of the 10th anniversary of the activity. We are looking forward! We are making efforts! We must achieve our goal!

Chen Jingyong
September, 2017

18日（北京市，北京建筑大学）北京民居考察、答辩评审会及表彰奖励仪式结束；由学会和北京建筑大学负责汇集参加高校、联合企业、讲座与评审专家等的书稿，主编出版《国匠承启卷——传统民居保护性利用设计》；11月8日—10日将在学会2017年第27届（江西）年会暨国际学术交流会上呈现专题展览。

《国匠承启卷》全书采用中英文对照方式编辑出版，将于第27届（江西）年会期间出版发行。将活动出版物作为中国建筑学会室内设计分会（CIID）与亚洲室内设计联合会（AIDIA）、国际室内建筑师／设计师联盟（IFI）等室内设计国际学术组织开展室内设计教育国际交流的特色案例之一，是本活动的一个拓展环节。

5卷的凝结。将活动作品集逐步打造成了中国建筑学会室内设计分会推荐教学参考书，也初步形成丛书的编制文化和装帧风格。每卷的6+1个"编制栏目"涵盖了相应"活动环节"的核心内容。其中，"调研踏勘""中期检查""答辩展示""教学研究""专家讲坛""历史定格"6个章节版块记录了当届活动的主体和过程，还附有专家点评、学生感言、获奖证书、师生照片等；经过5届不断完善、优化形成的《活动章程》（2017修订版）、本届活动纲要、《毕业设计框架任务书》《答辩、评奖、表彰工作细则》等，汇编在第1章"活动规章"中，是指导活动开展的长效机制，是学会室内设计教育工作有章可循的体现。

总之，希望编入的活动印记较多，限于篇幅，编写之中难免挂一漏万，敬请指正，以利改进。

响应国家创新驱动发展战略和服务经济社会发展需求，培养造就一大批创新能力强、适应需要的高质量室内设计师，始终是全社会等共同面对的命题。

CIID"室内设计6+1"2017（第五届）校企联合毕业设计活动已圆满结束。从校企联合毕业活动中走来的建筑学、工业设计、环境设计、艺术与科技等专业的毕业生们，已相应获得了建筑学学士（专业学位）、工学学士、艺术学学士学位，他们继而有的考取了硕士研究生，有的出国留学深造，有的已就业开始工作，共同踏上了面向行业需求、走向卓越发展的新征途。

今年是中国建筑学会室内设计分会成立28周年。在此，以《国匠承启卷》作为学会室内设计教育示范项目——"室内设计6+1"校企联合毕业设计活动创立5周年纪念版，呈献给朋友们。也希望通过学会各地方专业委员会的协同推动，能进一步组织开展区域性、特色性"室内设计6+1"校企联合毕业设计活动，进一步拓展品牌交流平台，提升我国室内设计人才培养质量。

感谢全国高等学校建筑学学科专业指导委员会、教育部高等学校设计学类专业教学委员会等长期以来对高校相关学科专业建设工作的指导！

感谢亚洲城市与建筑联盟（AAUA）和亚洲设计学年奖组委会、学会地方专业委员会、相关高校、命题企业、讲座与评审专家、出版企业等对CIID"室内设计6+1"校企联合毕业设计活动的支持和帮助！

感谢活动参加高校5年来的承办建设，以及甘当志愿者的朋友们！

2017/2018学年秋季学期开学了，2017级新生入校了。他们的作品或将刊载于活动10周年时的纪念版上。我们期待！我们努力！我们实现！

陈静勇
2017年9月

國匠承啓卷——傳統民居保護性利用設計

分会"室内设计 6+1"
式暨亚洲设计学年奖 2017 广州论坛

目 录

Preface
序 ·················· 005

Foreword
前言 ·················· 007

一 Activity Rules
活动规章 ·················· 015

CIID"室内设计6+1"2017（第五届）校企联合毕业设计活动纲要 ·················· 016
中国建筑学会室内设计分会（CIID）"室内设计6+1"校企联合毕业设计活动章程（2017修订版） ·················· 019
中国建筑学会"校企联合毕业设计6+"标志设计 ·················· 022
CIID"室内设计6+1"2017（第五届）校企联合毕业设计框架任务书 ·················· 025
CIID"室内设计6+1"校企联合毕业设计答辩、评奖、表彰工作细则（2017修订版） ·················· 027

二 Investigation and Survey
调研踏勘 ·················· 029

三 Medium-term Inspection
中期检查 ·················· 069

四 Defense Presentation
答辩展示 ·················· 095

榆林市赵庄景观规划与建筑保护性更新及再利用设计 ·················· 096
社区文化活动中心 ·················· 104
上海近代黑石公寓及其周边环境保护与更新设计 ·················· 108
情系黄土 重返家园——陕州地坑院的传承与保护性设计 ·················· 114
老城新生 ·················· 122
市-井-里-院——哈尔滨市道外区中华巴洛克传统民居里院街区改造设计 ·················· 130
半山村山野旅宿与竹韵茶楼设计 ·················· 138
上海近代黑石公寓及其周边环境保护与更新设计 ·················· 148
一房一社区 ·················· 156
里里·里院——哈尔滨老道外传统街区院落再利用与里院商业的探索 ·················· 164
青龙胡同酒店设计 ·················· 172
拾遗——南京市非物质文化遗产展 ·················· 180
围炉共笙——浙江省台州市黄岩区半山村传统营造技艺传承及展示设计 ·················· 188
匠心椅阑干 ·················· 196

五 Teaching Research
教学研究 ……………………………… 205

2017年毕业设计指导心得与思考
左琰 ………………………………………… 207

旧建筑改造设计课题中的前期调研初探
薛颖 ………………………………………… 211

毕业设计中技术理念教育的思考
周立军 马辉 兆翚 …………………………… 215

由"传统民居保护利用设计"引发的教学思考
谷秋琳 刘晓军 ……………………………… 219

北京青龙胡同街区有机更新初探
杨琳 朱宁克 刘璧凝 滕学荣 ………………… 223

空间构筑系列课题的实践与思考
吕勤智 宋扬 ………………………………… 229

六 Expert Rostrum
专家讲坛 ……………………………… 233

王世福 ……………………………………… 235
臧峰 ………………………………………… 235
谭善隆 ……………………………………… 237
王昀 ………………………………………… 239
王炜民 ……………………………………… 241
赵健 ………………………………………… 241

七 History Freeze-frame
历史定格 ……………………………… 243

活動規章
Activity Rules

CIID「室内设计 6+1」2017（第五届）
校企联合毕业设计
CIID"Interior Design 6+1"2017(Fifth Session)
University-Enterprise Cooperative Graduation Design

国匠承启卷
——传统民居保护性利用设计
Craftsmanship Heritage
——Design for the Protective Utilization of Traditional Folk Houses

壹

國匠承啓卷——傳統民居保護性利用設計

CIID "室内设计 6+1"
2017（第五届）校企联合毕业设计活动纲要

一、课　　题：国匠承启——传统民居保护性利用设计
二、项目地点：活动参加高校负责组织考察、遴选所在省市地域的代表性传统民居，建议优先选择国家或省市划定的传统民居类的历史文化街区（或历史文化保护区、历史文化名镇、历史文化名村等），作为在本届项目总体目标指导下的具体项目地点。
三、主办单位：中国建筑学会室内设计分会（CIID）
四、承办单位：中国建筑学会室内设计分会 2016 年第二十六届（杭州）年会组委会（命题研讨会）
　　　　　　　华南理工大学（开题仪式、广东民居考察）
　　　　　　　浙江工业大学（中期检查、浙江民居考察）
　　　　　　　北京建筑大学（答辩评审、北京民居考察、表彰奖励、《国匠承启卷》总编）
五、参加高校（学院＼专业）：
　　　　　　　同济大学（建筑与城市规划学院＼建筑学）
　　　　　　　华南理工大学（设计学院＼环境设计）
　　　　　　　哈尔滨工业大学（建筑学院＼建筑学、环境设计）
　　　　　　　西安建筑科技大学（艺术学院＼环境设计）
　　　　　　　北京建筑大学（建筑与城市规划学院＼环境设计）
　　　　　　　南京艺术学院（工业设计学院＼艺术与科技）
　　　　　　　浙江工业大学（艺术学院＼环境设计）
六、命题企业：由参加高校协同历史文化街区（或历史文化保护区、历史文化名镇、历史文化名村）保护规划编制单位或街区更新设计、开发企业等（一校一企业）
七、合作支持：亚洲城市与建筑联盟（AAUA）、亚洲设计学年奖组委会
八、出版企业：中国水利水电出版社
九、媒体支持：中国室内设计网 http://www.ciid.com.cn
十、时　　间：2016 年 10 月 － 2017 年 10 月
十一、活动安排：

序号	阶段	时间	地点	活动内容	相关工作
1	命题研讨	2016/10/29	学会二十六届（杭州）年会	• 10/27，报到，场外活动 • 10/28，学会二十六届（杭州）年会开幕式及学术交流活动 • 10/29，2017（第五届）活动命题研讨会	• 总结交流往届活动经验 • 研讨《2017（第五届）校企联合毕业设计框架任务书》 • 确定活动承办高校工作
2	教学准备	2016/10/30 － 2017/03/02	各高校	• 校企联合毕业设计教学工作准备	• 各高校报送参加毕业设计师生名单 • 各高校反馈对《活动纲要》的修改意见和建议 • 各高校邀请联合企业（一校一企业），依据《框架任务书》，分别编制和向学会报送本校《毕业设计详细任务书》 • 各高校安排文献检索与民居考察毕业实习 • 编制开题调研汇报 PPT、书稿开题调研分配页面电子版 • 准备开题仪式和民居考察
3	开题踏勘	2017/03/03 － 2017/03/05	华南理工大学	• 03/03，报到 • 03/04，开题仪式 　专题讲座 　调研交流 • 03/05，广东民居考察	• 开题仪式：学会颁发导师聘书 • 专题讲座：学会颁发讲座专家聘书 • 开题调研汇报；提交开题调研成果电子版 • 商讨中期检查、编辑出版等相关工作 • 民居考察

活動規章

序号	阶段	时间	地点	活动内容	相关工作
4	方案设计	2017/03/06 — 2017/04/13	各高校	● 方案设计	● 各高校安排相关讲授、考察、辅导、设计、研讨等 ● 选择课题方向，明确方案设计目标，完成相关文案、图表等 ● 完成调研报告、方案设计基本图示、效果图、模型、分析图表及等相应成果 ● 编制中期检查方案设计汇报PPT、书稿中期检查分配页面排版 ● 准备中期检查和民居考察
5	中期检查	2017/04/14 — 2017/04/16	浙江工业大学	● 04/14，报到 ● 04/15，中期检查 ● 04/16，浙江民居考察	● 中期检查汇报；提交中期检查成果电子版 ● 专家点评，形成方案设计深化重点 ● 商讨答辩评审、表彰奖励、编辑出版、展览等相关工作 ● 民居考察
6	深化设计	2017/04/17 — 2017/06/14	各高校	● 方案深化设计	● 完成方案深化设计图示、效果图、模型、分析图表、设计说明等相应成果；编制展板 ● 编制毕业设计答辩汇报PPT、书稿答辩方案分配页面排版 ● 准备答辩评审、民居考察、表彰奖励
7	答辩评审	2017/06/16 — 2017/06/18	北京建筑大学	● 06/16，报到 ● 06/17，答辩、评审 ● 06/18，北京民居考察；表彰奖励	● 答辩布展与观摩；提交书稿答辩方案成果电子版 ● 毕业答辩；等级奖评审 ● 学会表彰毕业设计等级奖、优秀毕业设计指导教师、毕业设计最佳组织单位、毕业设计突出贡献单位等 ● 2017（第五届）活动总结
8	编辑出版	2017/06/19 — 2017/10/25	学会、参加高校、命题企业、支持企业、出版企业	● 07/20前，高校完成《国匠承启卷》作品书稿部分编辑工作，提交学会秘书处 ● 09/15前，完成书稿总编工作，提交出版企业校审 ● 09/30前，完成书稿校审工作，形成清样，送印厂 ● 10/25前，《国匠承启卷》刊印完成	● 各高校和相关专家提交书稿其他页面内容 ● 学会和北京建筑大学负责总编《国匠承启卷》 ● 出版企业负责书稿审校出版 ● 学会二十七届（江西）年会2018（第六届）校企联合毕业设计命题研讨会准备
9	展览交流	2017/11/08 — 2017/11/10	学会二十七届（江西）年会室内设计教育论坛	● 11/08，报到 ● 11/09，《国匠承启卷》发行式暨2018（第六届）校企联合毕业设计命题研讨会 ● 11/08-11/10，学会年会专题展览	● 展板布展与观摩 ● 发行《国匠承启卷》 ● 研讨确定校企联合毕业设计2018（第六届）毕业设计命题 ● CIID"室内设计6+1"2017（第五届）校企联合毕业设计专题展览

Articles of Association of CIID "Interior Design 6+1" University-Enterprise Cooperative Graduation Design Program(Revised Edition 2017)

To meet training demand of specialized interior design talents for urban and rural construction, train interior designers more specifically, promote the exchanges in interior design specialization development and teaching among the colleges and universities, and unfold the graduation design teaching as per the demand of architectural industry (enterprises), the China Institute of Interior Design ("CIID") has initiated and supervised the College-Enterprise Cooperative Graduation Design Program that involves colleges and universities offering interior design related programs, industry representative architectural firms and interior design firms.

To standardize the Cooperative Graduation Design Program and grow it as a characteristic brand, CIID has prepared the Articles of Association in light of ideas and suggestions of relevant colleges and universities, deliberated and approved the same at the 1st CIID Interior Design 6+1 College-Enterprise Cooperative Graduation Design Topic Conference 2013 for trial implementation.

I. Program Background, Purpose and Significance

The Ministry of Education initiated the Excellent Engineer Education and Training Program in 2010, and released the undergraduate specializations and postgraduate disciplines under the Excellent Program in three batches from 2011 to 2013. As shown in the Directory of Degree Awarding and Talent Training Disciplines (2011) jointly announced by the Degree Committee and Ministry of Education under the State Council in 2011, the discipline category of Arts (13) is included and Design (1305) is rated a first-level discipline under the discipline category of Arts; Environmental Design is suggested as a second-level discipline under the first-level discipline of Design; Interior Design is proposed as a second-level discipline under the newly adjusted first-level discipline of Architecture (0813). As revealed in the Directory of Undergraduate Specializations of Regular Institutions of Higher Education (2012) released by the Ministry of Education in 2012, Design (1305) is set as a specialization under the discipline category of Arts, and Environmental Design (130503) is among the core specializations under Design. The independent setting of Arts as a discipline category, the setting and adjustment of Design as a first-level discipline, and Environmental Design, Interior Design and other specializations have formed a new pattern of China's environmental design education and interior design talent training disciplines and specializations.

The College-Enterprise Cooperative Graduation Design Program is of great significance to strengthening characteristic development of relevant disciplines and specializations in the implementation of the Excellent Engineer Education and Training Program initiated by the Ministry of Education, deepening exchanges in teaching process of graduation design, promoting synergic innovation of interior design education and teaching, and training interior design talents to meet industry (enterprise) demands.

II. Program Organizational Structure

1. Sponsor

The Program is sponsored by CIID or a CIID authorized local specialized committee ("Specialized Committee") under the guidance and support of National Steering Committee of Architecture Education in China, Steering Committee of Design Education of the Ministry of Education and other committees.

2. Participating Colleges and Organizing Colleges

Upon consultation, 6 colleges with similar discipline and specialization conditions offering interior design related specializations will be selected as participating colleges. The representativeness in geographical region, type of college operation, specialization features and employment orientation of the participating colleges should be highlighted to form an interdisciplinary and synergic design environment.

Every year 3 colleges are selected from the participating colleges as organizing colleges for proposal survey, mid-term examination, oral defense review and commendation, editing and publication of Album under the Theme of X—General Topic of X [CIID Recommended Textbook: The XX CIID Interior Design 6+1 College-Enterprise Cooperative Graduation Design Program 20XX] ("Theme Album") respectively.

Each participating college is expected to send 6 students divided into 2 design teams for the oral defense. The students should be accompanied by 1 to 2 college mentors, at least one of whom should hold a senior title, know environment design, interior design and other engineering practice well, and have extensive connections with enterprises in the related field.

3. Topic Selecting Enterprise

The college organizing the proposal survey of the Program recommends to CIID or the Specialized Committee an industry representative architectural and interior design firm incorporated in the province (city) where it's located as the general topic selecting enterprise. The general topic selector should hold a senior specialized technical post and will often serve as the General Mentor and Leader of the Judge Panel. Headed by the General Mentor, the general topic selecting enterprise is to lead the preparation of The Framework Specifications under the Theme of X—General Topic of X for College-Enterprise Cooperative Graduation Design Program, and participate in the proposal survey, mid-term examination, oral defense review and commendation, album editing and publication, themed exhibition and external exchange.

4. Supporting Enterprises

Every year participating colleges recommend to CIID or the Specialized Committee industry representative architectural and interior design firms as supporting enterprises of graduation design. CIID or the Specialized Committee signs Support and Feedback Agreement with the supporting enterprises to involve them in the graduation design. CIID or the Specialized Committee requests the supporting enterprises to select experts for seminars based on the general topic of graduation design, and for proposal survey, mid-term examination, oral defense review and commendation, album editing and publication, themed exhibition and external exchange.

5. Publishing Enterprises

CIID or the Specialized Committee selects an industry representative publishing enterprise as the publisher of the Theme Album of the year as per the College-Enterprise Cooperative Graduation Design Program.

III. Organizational Flow

(1) The College-Enterprise Cooperative Graduation Design Program is held annually as per the teaching arrangement of graduation design of colleges and universities in respect of the graduates.

(2) The College-Enterprise Cooperative Graduation Design Program is primarily composed of 6 parts: graduation design topic seminar, proposal survey, mid-term examination, oral defense review and commendation, Theme Album editing and publication,

中国建筑学会室内设计分会（CIID）"室内设计6+1"校企联合毕业设计活动章程（2017修订版）

为服务城乡建设领域室内设计专门人才培养需求，加强室内设计师培养的针对性，促进相关高等学校在室内设计学科专业建设和教育教学方面的交流，引导相关专业面向建筑行（企）业需求组织开展毕业设计教学工作，由中国建筑学会室内设计分会（CIID，以下简称"学会"）倡导、主管，国内设置室内设计相关学科专业的高校与行业代表性建筑与室内设计企业开展联合毕业设计。

为使联合毕业设计活动规范、有序，形成活动品牌和特色，学会在征求相关高等学校意见和建议的基础上形成本章程，并于学会CIID"室内设计6+1"2013（首届）校企联合毕业设计命题会上审议通过，公布试行。

一、校企联合毕业设计活动设立的背景、目的和意义

2010年教育部启动了"卓越工程师教育培养计划"，于2011-2013年分三批公布了进入"卓越计划"的本科专业和研究生层次学科。2011年国务院学位委员会、教育部公布了《学位授予和人才培养学科目录（2011年）》，增设了"艺术学（13）"学科门类，将"设计学（1305）"设置为"艺术学"学科门类中的一级学科。"环境设计"建议作为"设计学"一级学科下的二级学科，"室内设计"建议作为新调整的"建筑学（0813）"一级学科下的二级学科。2012年教育部公布了《普通高等学校本科专业目录（2012年）》，在"艺术学"学科门类下设"设计学类（1305）"专业，"环境设计（130503）"等成为其下核心专业。"艺术学"门类的独立设置，设计学一级学科以及环境设计、室内设计等学科专业的设置与调整，形成了我国环境设计教育和室内设计专门人才培养学科专业的新格局。

举办室内设计校企联合毕业设计活动，对在教育部"卓越工程师教育培养计划"实施中加强相关学科专业特色建设，深化毕业设计各教学环节交流，促进室内设计教育教学协同创新，培养服务行（企）业需求的室内设计专门人才，具有十分重要的意义。

二、校企联合毕业设计活动组织机构

1. 主办单位

校企联合毕业设计活动由学会或经授权的地方专业委员会（以下简称专委会）主办，得到了全国高等学校建筑学学科专业指导委员会、教育部高等学校设计学类专业教学指导委员会等的指导和支持。

2. 参加高校、承办高校

校企联合毕业设计活动一般由学科专业条件相近，设置室内设计方向的相关专业的6所高校间通过协商、组织成为活动参加高校。应突出参加高校的地理区域、办学类型、专业特色、就业面向等的代表性，在学科专业间形成一定的交叉性和协同设计环境。

每年在参加高校中遴选其中3所高校分别作为毕业设计开题调研、中期检查、答辩评审与表彰奖励、《主题×卷——总命题×》[中国建筑学会室内设计分会推荐教学指导书 CIID"室内设计6+1"×年（第×届）校企联合毕业设计]（以下简称《主题卷》）总编出版等活动的承办高校。

每所高校参加联合毕业设计到场汇报的学生一般以6人为宜，分为2个设计方案组；要求配备1～2名指导教师，其中至少有1名指导教师具有高级职称，熟悉环境设计、室内设计等工程实践业务，与相关领域企业联系较广泛。

3. 命题企业

承办校企联合毕业设计开题调研的高校，负责向学会或专委会推荐所在省（市）的行业代表性建筑与室内设计企业作为毕业设计总命题企业，总命题人具有高级专业技术职务，一般作为联合毕业设计总导师和评审专家组组长。总命题企业以总导师为主，负责牵头编制《"主题×——总命题×"联合毕业设计框架任务书》，参与开题调研、中期检查、答辩评审与表彰奖励、总编出版、专题展览、对外交流等工作。

4. 支持企业

参加高校在每届校企联合毕业设计活动中，分别向学会或专委会推荐行业代表性建筑与室内设计企业作为毕业设计支持企业，由学会或专委会与支持企业签订活动支持与回馈协议，安排支持企业参与毕业设计活动。学会或专委会负责联系支持企业选派专家，围绕毕业设计总命题举办专题讲座，参与毕业设计开题调研、中期检查、答辩评审与表彰奖励、编辑出版、专题展览、对外交流等活动。

5. 出版企业

学会或专委会就每届校企联合毕业设计活动，遴选行业代表性出版企业承担当届《主题卷》的出版工作。

三、校企联合毕业设计活动组织流程

（1）校企联合毕业设计活动按照参加高校毕业设计教学工作安排在每个年度（按毕业生届次）举行1次。

（2）校企联合毕业设计活动主要教学环节包括毕业设计命题研讨、开题调研、中期检查、答辩评审与表彰奖励、《主题卷》编辑出版、专题展览6个主要环节，对外交流作为联合毕业设计活动的1个扩展环节。

（3）学会或专委会作为活动主办单位，负责活动总体策划、宣传，协调参加高校、相关企业、展览机构等，聘请有关专家举办专题学术讲座，组织毕业设计学年奖、毕业设计优秀指导教师、毕业设计优秀组织单位、毕业设计特殊贡献奖等的评选、表彰，以及室内设计教育国际交流工作。

（4）参加高校在学会或专委会的指导、协调下，联合相关企业等，共同拟定校企联合毕业设计活动纲要，向学会备案。联合毕业设计开题调研、中期检查、答辩评审与表彰奖励、《主题卷》总编出版等工作由承办高校分工落实。

（5）命题研讨。参加高校的毕业设计指导教师参加联合毕业设计命题研讨会。毕业设计总命题由命题企业与参加高校着眼文化遗产保护、城市设计、建筑设计、室内设计、展示设计、产品设计等领域发展前沿和行业热点问题，结合参加高校毕业设计教学实际商讨形成，报学会备案。毕业设计命题要求具备相关资料收集和现场踏勘等条件。命题研讨会一般安排在高校秋季学期中（每年11月左右），结合当年学会年会安排专题研讨。

（6）开题仪式与现场踏勘。参加高校的毕业设计师生参加联合毕业设计开题调研，命题企业提供毕业设计必要的设计基础资料和设计任务需求，指导现场踏勘等活动等。开题活动一般安排在高校春季学期开学初（3月上旬）进行。

（7）中期检查。参加高校的毕业设计师生参加联合毕业设计中期检查活动。参加高校毕业设计指导教师和相关企业专家等，对毕业设计中期成果进行检查、评审，开展教学交流。中期检查一般安排在春季学期期中（4月下旬）进行。

中期检查活动中，每所参加高校优选不超过2个方案组进行陈述与答辩；其中陈述不超过10分钟，问答不超过10分钟。

（8）答辩评审与表彰奖励。参加高校的毕业设计师生参加联合毕业设计答辩评审与表彰奖励活动。学会或专

and themed exhibition. External exchange is an extended part of the Program.

(3) As the sponsor, CIID or the Specialized Committee is responsible for overall planning and promotion of the Program, coordination among participating colleges, related enterprises and exhibition agencies, engagement of related experts to hold themed academic lectures, organization of appraisal and commendation of Graduation Design Academic Year Award, Excellent Mentor of Graduation Designs, Excellent Organizer of Graduation Design and Special Contributor of Graduation Design, and international exchanges in interior design education.

(4) As guided and coordinated by CIID or the Specialized Committee, participating colleges work with related enterprises to map out the Program Outline and file it with CIID. Proposal survey, mid-term examination, oral defense review and commendation, Theme Album editing and publication are to be undertaken by the organizing colleges.

(5) Topic Seminar. The graduation design mentors of participating colleges attend the graduation design topic seminar. The general topic of graduation design is defined by the topic selecting enterprise and participating colleges in regard to development trends and hot issues of cultural heritage protection, urban design, architectural design, environmental design, interior design, display design and product design in light of discussions on teaching practices of graduation design of participating colleges. It's to be filed with CIID. The graduation design topic should meet the requirements on related design information collection and site survey. The topic seminar is often held in the middle of autumn semester (around November) in combination of the CIID Annual Conference of the year.

(6) Proposal Ceremony and Site Survey. The graduation design teachers and students of participating colleges take part in the proposal survey of graduation design. The topic selecting enterprise provides basic design information and design task requirements essential for graduation design, and guides the site survey. Proposal related activities are generally held at the beginning of spring semester (in early March).

(7) Mid-term Examination. The graduation design teachers and students of participating colleges take part in the mid-term examination of graduation design. The college mentors of graduation design and related enterprise experts examine and review the mid-term results of graduation design and make teaching exchanges. The mid-term examination is usually unfolded in the middle of spring semester (in late April).

During the mid-term examination, each participating college selects no more than 2 design teams for presentation and oral defense both to be completed within 10 minutes.

(8) Oral Defense Review and Commendation. The graduation design teachers and students of participating colleges participate in the oral defense review and commendation of graduation design. The experts of CIID or the Specialized Committee, experts of the topic selecting enterprise and college mentors of graduation design take part in the oral defense review and commendation of graduation design as judges. The oral defense review and commendation of graduation design is often unfolded at the end of spring semester (in early June).

For the graduation oral defense, each participating college selects no more than 2 design teams for presentation and oral defense both to be completed within 15 minutes, and for result exhibition. Furthermore, each participating college may select no more than 2 self recommendation plans for result exhibition, but not for presentation and oral defense.

After the oral defense, result exhibition and review, CIID or the Specialized Committee organizes the appraisal of Graduation Design Awards, Excellent Mentor of Graduation Design, Excellent Organizer of Graduation Design and Special Contributor of Graduation Design, and awards the winners. The review principles of CIID reputation and quality first, and quality before quantity are to be upheld. Graduation Design Awards are categorized into the First Place Award, Second Place Award and Third Place Award in the ratio of 1:2:4, and Excellence Award. The Rank Awards are to be awarded to oral defense plans only and the Excellence Award to be awarded to self recommendation plans exclusively.

(9) Themed Exhibition. Upon conclusion of the Program every year, CIID or the Specialized Committee arranges a themed exhibition of the graduation designs at the CIID Annual Conference (to be convened from October to November each year). After the themed exhibition, related colleges may apply with CIID or the Specialized Committee for an exhibition tour of graduation designs.

(10) Editing and Publication. On the basis of the Cooperative Graduation Design Program of the year, CIID or the Specialized Committee organizes the editing and publication of the Theme Album as CIID recommended textbook. The Theme Album is to be chiefly edited by CIID or the Specialized Committee and corresponding organizing colleges; the participating colleges work as co-editors and the college mentors of graduation design are responsible for reviewing the typeset manuscript of respective college; the publishing enterprise serves as the editor in charge who is responsible for proofreading, publication, distribution and promotion.

(11) External Exchange. Generally, the Release Ceremony of Theme Album is held at the CIID Annual Conference of the year upon conclusion of the Program. CIID invites the Asia Interior Design Institute Association (AIDIA) and International Federation of Interior Architects/Designers (IFI) for an international exchange in educational results of interior design to expand and publicize China's interior design education and diversify the approaches of international exchanges.

IV. Funds

(1) CIID or the Specialized Committee raises funds for presenting Graduation Design Awards, Excellent Mentor of Graduation Design, Excellent Organizer of Graduation Design and Special Contributor of Graduation Design, and organizing themed exhibition at the CIID Annual Conference.

(2) Participating colleges are responsible for the travelling expenses of personnel participating the Program, venue fees and related expenses for exhibition of graduation designs of the college.

(3) Organizing colleges bear the venue fees, conference expenses and organization costs in relation to proposal survey, mid-term examination, oral defense review and commendation, and publication fees of Theme Album.

(4) The topic selecting enterprise, supporting enterprises and the publishing enterprise subsidize the Program in a certain form.

V. Supplementary Provisions

(1) The Articles of Association were issued on and have come into effect as of January 13, 2013, the right of interpretation of which rest in CIID.

(2) The Articles of Association were first revised in March 2014, revised for the second time in April 2015, and last revised in June 2017.

委会专家、命题企业专家、参加高校毕业设计指导教师等作为评委，参加毕业设计答辩评审、表彰奖励等工作。毕业设计答辩评审与表彰奖励一般安排在春季学期期末（6月上旬）进行。

毕业答辩活动中，每所参加高校优选不超过2个答辩方案组进行陈述与答辩、成果展出；其中陈述不超过15分钟，问答不超过15分钟。此外，每所参加高校可再安排不超过2个自荐方案组进行成果展示，不参加陈述与答辩。

在答辩、成果展示、评审的基础上，学会或专委会组织评选毕业设计奖、毕业设计优秀指导教师、毕业设计优秀组织单位、毕业设计特殊贡献奖等，并给予表彰奖励。坚持学会声誉和质量第一、宁缺毋滥的评审原则。毕业设计奖设一、二、三等奖、优秀奖等。其中，一、二、三等奖一般按照1:2:4比例设置；等级奖评选仅针对答辩方案设置，优秀奖针对自荐方案设置。

（9）专题展览。学会或专委会在每届联合毕业设计结束当年学会年会（每年的10月—11月）上安排毕业设计作品专题展览；年会专题展览结束后，相关高校可向学会或专委会申请毕业设计作品巡回展出。

（10）编辑出版。基于每届联合毕业设计活动，由学会或专委会组织编辑出版《主题卷》，作为学会推荐的教学参考书。《主题卷》总编工作由学会或专委会和相应承办高校联合编辑；参加高校作为参编单位，参加高校毕业设计指导教师负责本校排版稿的审稿等工作；出版企业负责出版发行等工作。

（11）对外交流。一般在当届联合毕业设计结束当年学会年会上举行《主题卷》发行式，并由学会联系亚洲室内设计联合会（AIDIA）、国际室内建筑师／设计师联盟（IFI）等，开展室内设计教育成果国际交流，拓展宣传中国室内设计教育，拓展国际交流途径。

四、校企联合毕业设计活动经费

（1）学会或专委会负责筹措评选毕业设计奖、毕业设计优秀指导教师、毕业设计优秀组织单位、毕业设计特殊贡献奖等表彰奖励经费，以及学会年会专题展览经费等。

（2）参加高校自筹参加校企联合毕业设计活动相关人员的差旅费，以及毕业设计作品在本校巡展的场地及相关经费等。

（3）承办高校自筹校企联合毕业设计开题调研、中期检查、答辩评审与表彰奖励等相关活动的场地、会议费、组织费，以及《主题卷》出版经费等。

（4）命题企业、支持企业、出版企业等负责为向校企联合毕业设计活动提供一定形式的资助等。

五、附则

（1）本章程2013年1月13日通过并公布施行，由中国建筑学会室内设计分会负责解释。

（2）本章程2014年3月第一次修订；2015年4月第二次修订；2017年6月第四次修订。

中国建筑学会室内设计分会"校企联合毕业设计6+"标志设计

林则全[1]、孙小鹏[1]、叶红[2]、陈静勇[1T]

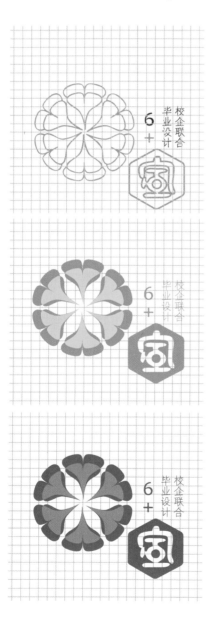

三十年著子，三千年沧桑。作银杏以为梁，饰以时间，阅尽古今。西汉《上林赋》述"上千韧，达连抱，夸条直畅，实叶茂盛"，可称端直葱茏，坚牢庄重。

白驹过隙，已然千年。夫物芸芸，落叶满黄。CIID"室内设计6+1"校企联合毕业设计教育项目自2013年创立以来，围绕人才培养，开展五届，现已枝干初成，由此托生出互联协同的"校企联合毕业设计6+"新项目。今"校企联合毕业设计6+"，师银杏生长精神，承黄叶广博茂盛，以此为基，代代相传。

"校企联合毕业设计6+"新标志基本形由银杏叶组成，六瓣一轮，佐以原CIID六角"室"字徽，传承学会主旨精神，引出教育项目内涵。轮状散叶，生出朵朵金花，满目绚烂。平面形态似阳光普照，有如百花绽放，喻新生力量源远流长。

标志基本形由银杏叶、六角、圆、加号等组成，寓意人才培养、学会平台、规矩制度、协同创新的广泛联合。

主体色中，黄源自CIID标志六角"室"字，色为银杏秋色，其态有力；蓝继原"室内设计6+1"校企联合毕业设计活动标志，稳重典范。黄出创意源泉，蓝系规范标准，二色对照，相辅相成，体现服务行业人才培养需求。

辅助色提取红、绿、白、灰、黑作为辅助色，其五色配合主体色，共七色，作为新标志色彩。红立宫墙，现典雅、华美；绿源彩画，表清新、鲜活；白出黎明，预新生、期盼；灰随城垣，示坚实、守护；黑为燃尽百草，寓涅槃、重生。

拓展色为主体色和辅助色的滤色叠合产生，更加鲜亮，更为活泼。

---

1 北京建筑大学
2 中国建筑学会室内设计分会
T 通讯作者

基本图形

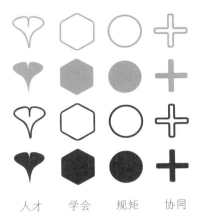

人才　学会　规矩　协同

基本字体

方正兰亭黑
ABCDEFGHIJKLMNOPQRSTUVWXYZ
abcdefghijklmnopqrstuvwxyz

汉仪家书简
ABCDEFGHIJKLMNOPQRSTUVWXYZ
abcdefghijklmnopqrstuvwxyz

拓展字体

康熙字典体
方正清刻本悦宋简体
Cataneo BT

主体色

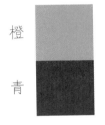

橙　C18 M37 Y83 K0
　　R214 G167 B59

青　C96 M88 Y53 K25
　　R22 G46 B78

辅助色

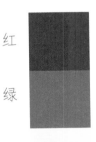

红　C46 M100 Y99 K17
　　R139 G28 B33

绿　C80 M42 Y59 K1
　　R52 G123 B112

白　C1 M9 Y9 K0
　　R251 G238 B230

灰　C63 M53 Y51 K1
　　R114 G116 B115

黑　C80 M75 Y73 K49
　　R46 G46 B46

拓展色

活動規章

The Framework Specifications for CIID "Interior Design 6+1" 2017 (Fifth Session) University-Enterprise Cooperative Graduation Design Program

The Framework Specifications for CIID "Interior Design 6+1" 2017 (Fifth Session) University-Enterprise Cooperative Graduation Design Program have been prepared by the college mentors of CIID "Interior Design 6+1" University-Enterprise Cooperative Graduation Design Program based on the results of the 26th CIID (Hangzhou) Annual Meeting for Topic Seminar 2016. Participating colleges should work with the relevant well-known design firm to further prepare The Detailed Specifications for Graduation Design in respect of the college's teaching practice of graduation design to guide the teaching of graduation design and unfold the graduation design activities.

I. Program Name

Lineage of Craftsmanship—Protective Utilization Design of Traditional Folk Houses
(Sub-Program Name to be decided by respective colleges)

II. General Principles and Goals

(1) Protective Utilization Design Principles of Traditional Folk Houses: suitability to living, business, tourism and inheritance.

(2) Development Goals of Protective Utilization of Traditional Folk Houses: Allow the elderly to enjoy their late years locally, draw their children back for hometown development, carry forward construction skills and strengthen the history and culture.

III. Program Site

Participating colleges are responsible for organizing surveys to and selecting the representative traditional folk houses in the region, city and province where the college is located as the program site under the general goals, preferably historical blocks (or historic and cultural protected areas, famous historic and cultural towns and famous historic and cultural villages) categorized as traditional folk houses by the city, province or state.

IV. Scope of Design

Participating colleges work with preparing units of conservation planning of historical blocks (or historic and cultural protected areas, famous historic and cultural towns and famous historic and cultural villages), block renewal design and development enterprises to define the scope of design within the historical blocks (or historic and cultural protected areas, famous historic and cultural towns and famous historic and cultural villages) in light of the teaching practices of graduation design of respective college.

V. Topic Selection

In reference to the general goals of the Program of the year and self-defined scope of design, each participating college should instruct each design team of the college to select one of the following topics, and specify the key points of respective topic at The Detailed Specifications for Graduation Design:

1. Conservation Planning and Landscape Design of Folk Houses

Explore the conservation planning, architectural exterior landscape and site design methods of traditional folk houses on the basis of conservation planning of historical blocks (or historic and cultural protected areas, famous historic and cultural towns and famous historic and cultural villages).

2. Architectural Conservation and Renewal Design of Folk Houses

Explore the architectural conservation and interior and exterior renewal design methods of folk houses on the basis of street layout of traditional folk houses, spatial structure of settlements, architectural forms and colors, human geography and folk traditions, and development demands of social life within the scope of planning and design.

3. Inheritance and Exhibition Design of Construction Skills

Study the inheritance and exhibition design methods of construction skills in an active settlement environment on the basis of development history of construction skills of traditional folk houses, inheritors of Master Craftsmen of Traditional Architecture, construction skill structure of traditional folk houses, and construction mode lineage.

VI. Depth of Design

(1) Proposal Survey Report: Folk house survey report, design concept analysis, typeset proposal survey manuscript.
(2) Mid-term Examination Report: Preliminary design plan, typeset mid-term examination manuscript.
(3) Graduation Oral Defense Report: Detailed design plan, exhibition boards and typeset graduation oral defense manuscript.

VII. Design Results

(I) Design Notes

Design notes mainly consist of: elaboration of overall planning and design philosophy, planning and design orientation, design derivative process analysis, functional analysis, economic and technical indexes in graphical representation and diagram.

(II) Drawings (In Respect of Topic)

(1) Location map.
(2) Conservation plan and planning analysis diagram.
(3) Landscape plan, site plan and analysis diagram.
(4) Representative architecture floor plan, elevation, profile and analysis diagram.
(5) Interior floor plan, top view and elevation profile of building, detailed floor plan and analysis diagram of representative rooms.
(6) Furnishings drawing, sign system drawing and analysis diagram of representative rooms.
(7) Exhibition design drawing, detailed drawing and analysis diagram of exhibition facilities.
(8) Detailed drawing of key nodes.
(9) Main material list, furniture and furnishings list and other charts.

(III) Color Design Sketches

(1) Aerial view plan.
(2) Landscape design sketch.
(3) Architectural design sketch.
(4) Interior design sketch.
(5) Design sketch of environmental facilities, exhibition facilities and public art.

(IV) Result Submission

1. Proposal Exchange

Each participating college summarizes 1 report document for presentation and oral defense to be completed within 20 minutes (10 minutes for presentation and 10 minutes for oral defense), and submits proposal survey result PPT and manuscript.

2. Mid-term Examination

Each participating college selects 2 design teams for presentation and oral defense to be completed within 20 minutes (10 minutes for presentation and 10 minutes for oral defense), and submits mid-term examination result PPT and manuscript.

3. Graduation Oral Defense

(1) Each participating college selects 2 design teams for presentation and oral defense to be completed within 30 minutes (15 minutes for presentation and 15 minutes for oral defense), and submits graduation oral defense result PPT, exhibition boards and manuscript.

(2) Each design team prepares 3 exhibition boards on the basis of the template issued by CIID (A0 Extended Format: 900mm×1800mm vertical layout; resolution≥100dpi). The organizing college of graduation design oral defense is responsible for painting and exhibition arrangement.

4. Publication Materials

To edit and publish the Lineage of Craftsmanship Album—Protective Utilization Design of Traditional Folk Houses respectively (CIID Recommended Textbook: The Fifth CIID Interior Design 6+1 College-Enterprise Cooperative Graduation Design Program 2017), all units and persons participating in the Program should provide, as required by CIID, a unit introduction (within 1,000 Chinese characters), a teaching research paper (within 3,000 Chinese characters to be jointly written by mentors of each college), oral defense designs (each design of 4 or 6 pages in the topic of specialized design), summaries of expert lectures (1 from each expert within 1,500 Chinese characters), words from students (each design team within 200 Chinese characters), working photos (1 photo from each expert, mentor and student), award certificates (of all grades and types), photos taken during (the major processes of) the Program and other publishing materials in electronic form.

VIII. Attached Architectural Drawing and Site Map

See the Detailed Specifications of Respective College.

CIID "室内设计 6+1"
2017（第五届）校企联合毕业设计框架任务书

《CIID"室内设计6+1"2017（第五届）校企联合毕业设计框架任务书》是依据中国建筑学会室内设计分会（CIID）2016第二十六届（杭州）年会命题研讨会意见，经2017（第五届）校企联合毕业设计指导教师联合编制形成。活动参加高校应结合本校毕业设计教学工作实际，协同相关知名设计企业，据此进一步编制相应的《毕业设计详细任务书》，指导毕业设计教学工作，开展联合毕业设计活动。

一、项目名称
国匠承启——传统民居保护性利用设计
（子项名称由参加高校自拟）

二、总体原则和目标
（1）传统民居保护性利用设计原则。宜居，宜商，宜游，宜承。
（2）传统民居保护性利用发展目标。让长者就地能养老，让子女归巢能发展，让营造技艺能传承，让历史文化能衍生。

三、项目地点
活动参加高校负责组织考察、遴选所在省市地域的代表性传统民居，建议优先选择国家或省市划定的传统民居类的历史文化街区（或历史文化保护区、历史文化名镇、历史文化名村等），作为在本届项目总体目标指导下的具体项目地点。

四、设计范围
由参加高校协同历史文化街区（或历史文化保护区、历史文化名镇、历史文化名村等）保护规划编制单位和街区更新设计、开发企业等，结合本校毕业设计教学工作实际，在历史文化街区（或历史文化保护区、历史文化名镇、历史文化名村等）中划定出具体设计范围。

五、课题方向
参加高校基于本届联合毕业设计项目总体目标和自划定的设计范围，指导本校每个设计组选择以下课题方向之一，并在《毕业设计详细任务书》明确课题方向具体要点。

1. 民居保护规划与景观设计
基于历史文化街区（或历史文化保护区、历史文化名镇、历史文化名村等）保护规划，探讨规划设计范围内传统民居保护规划与建筑外环境景观及场地设计方式。

2. 民居建筑保护与更新设计
基于规划设计范围内的传统民居街巷布局、聚落空间肌理、建筑形态与色彩风貌、人文地理和民俗传统、社会生活发展需求等，探讨传统民居建筑保护与室内外更新设计方式。

3. 营造技艺传承与展陈设计
基于传统民居营造技艺发展历史沿革、"传统建筑名匠"传承人、传统民居营造技艺流程、构造方式谱系等，利用聚落活态环境，探讨有利营造技艺传承方式的展陈设计方式。

六、设计深度
（1）开题调研汇报：民居调研报告、设计概念分析、书稿开题调研内容排版页。
（2）中期检查汇报：设计初步方案、书稿中期检查内容排版页。
（3）毕业答辩汇报：设计深化方案、展板、书稿毕业答辩内容排版页。

七、设计成果
（一）设计说明
设计说明内容主要包含：整体规划与设计思想的阐述、规划与设计定位、设计衍生过程分析、功能分析、经济技术指标等，可作图示及图表。

（二）图纸（对应课题方向）
（1）区域位置图。
（2）保护规划图、规划分析图。
（3）景观规划设计图、场地设计图、分析图。
（4）代表性建筑平面图、立面图、剖面图、分析图。
（5）建筑室内平面图、顶面图、剖立面图、代表性房间平面详图、分析图。
（6）代表性房间陈设布置图、导识系统布置图、分析图。
（7）展陈设计布置图、展具设施详图、分析图。
（8）重要节点详图。
（9）主要材料表、家具陈设清单等图表。

（三）彩色效果图
（1）规划鸟瞰图。
（2）景观效果图。
（3）建筑效果图。
（4）室内效果图。
（5）环境设施、展具、公共艺术等效果图。

（四）成果提交
1．开题交流
每所参加高校汇总1个汇报文件进行陈述，每校答辩时间限20分钟（含陈述10分钟、问答10分钟）；提交开题调研成果PPT、开题调研内容书稿等。

2．中期检查
每所参加高校优选2个方案组进行陈述，每组答辩时间限20分钟（含陈述10分钟、问答10分钟）；提交中期检查成果PPT、中期检查内容书稿等。

3．毕业答辩
（1）每所参加高校优选2个答辩方案组进行陈述，每组答辩时间限30分钟（含陈述15分钟、问答15分钟）；提交毕业答辩成果、展板、毕业答辩内容书稿等。
（2）每个方案设计组的展板限3张，展板幅面A0加长：900mm×1800mm竖版，分辨率不小于100dpi。展板模板由学会按照年会展板要求统一提供。展览由毕业设计答辩活动承办高校负责喷绘、布置。

4．出版素材
为编辑出版《国匠承启卷——传统民居保护性利用设计》[中国建筑学会室内设计分会推荐教学参考书CIID"室内设计6+1"2017（第五届）校企联合毕业设计]，活动相关参加单位和个人等需积极响应学会要求，负责提供相应的单位简介（中文1000字以内）、教研论文（每所高校指导教师联写1篇，中文3000字以内）、答辩作品（每个方案在一个专业设计专题占4P或6P）、专家讲座提要（每位专家1篇，中文1500字以内）、专家寄语（每位专家中文300字左右）、专家点评（每位专家、导师对每个方案点评中文300字以内）、学生感言（每个方案组中文200字以内）、工作照片（每位专家、导师、学生各1张）、奖励证书（各级各类）、活动照片（各主要环节）等出版素材的电子文档。

八、附建筑与场地图
见各校详细任务书。

Oral Defense, Appraisal and Commendation Rules of CIID "Interior Design 6+1" University-Enterprise Cooperative Graduation Design Program(Revised Edition 2017)

The Oral Defense, Appraisal and Commendation Rules of CIID "Interior Design 6+1" University-Enterprise Cooperative Graduation Design Program have been formulated by CIID on the basis of The Articles of Association of CIID Interior Design 6+1 College-Enterprise Cooperative Graduation Design Program for work guidance, compliance and execution.

I. Preparation of Oral Defense

(1) Each college should send 2 mentors at most and 6 to 8 students for oral defense.

(2) Each college should arrange no more than 2 graduation design oral defense teams for graduation design oral defense, appraisal and exhibition of Graduation Design Rank Awards. Additionally 2 self recommendation plans can be submitted at most to CIID for appraisal and exhibition of Graduation Design Excellence Award, but not for graduation design oral defense, appraisal and exhibition of Graduation Design Rank Awards.

(3) Each graduation design oral defense team should prepare the graduation design oral defense PPT and other electronic documents in advance, and submit them to the Organizing Committee upon sign-in for graduation design oral defense.

(4) Before the oral defense, 3 exhibition boards should be prepared for each graduation design oral defense plan and self recommendation plan on the basis of the template issued by CIID (A0 Extended Format: 900mm×1800mm; resolution≥100dpi). the electronic exhibition boards must be sent to the work email address designated by the college organizing the oral defense 1 week before the oral defense. The college organizing the oral defense will be responsible for summarizing and printing out the exhibition boards for exhibition arrangement.

(5) The colleges should edit the manuscripts as per the Manuscript Typesetting Requirements for CIID "Interior Design 6+1" University-Enterprise Cooperative Graduation Design Program, and submit the same to the Organizing Committee at the graduation oral defense.

(6) Graduation Design Topic Selection and Submission . The colleges should select one topic for the 2 graduation design oral defense teams from the topics specified in The Framework Specifications under the Theme of X—General Topic of X for The "CIID Interior Design 6+1" University-Enterprise Cooperative Graduation Design Program of the year ("The Framework Specifications"), and submit the same to CIID 1 week before the oral defense.

II. Oral Defense and Appraisal

1. Oral Defense Judge Panel Composed of Special Judges and College Judges

(1) Special judges are 5 to 7 experts from topic selecting enterprise and CIID (including local specialized committees), program observers and enterprise supporting experts; the leader of the Oral Defense Judge Panel is usually the expert of the general topic selecting enterprise (General Mentor).

(2) College judges are graduate mentors elected from the colleges (1 mentor from each college).

2. Appraisal of Graduation Design Rank Awards

(1) The Graduation Design Rank Awards are categorized into the First Place Award, Second Place Award and Third Place Award in the ratio of 1:2:4 in accordance with the graduation design topic as decided in The Framework Specifications of the year; the awards may be vacant.

(2) The first round of appraisal. Each graduation design oral defense team defends their plan based on the selected topic within 15 minutes and answers the questions within 15 minutes. The special judges and college judges jointly fill in and sequence the ballot tickets (1 is the recommended order 1; 2 is the recommended order 2, so on and so forth.) The Organizing Committee counts the ballot tickets and sequences the design topics for appraisal of Graduation Design Rank Awards.

(3) The second round of appraisal. The special judges review and appraise the oral defense plans based on the recommended orders for appraisal of Graduation Design Rank Awards, and decide the Award-winning Plan for Graduation Design Rank Awards.

3. Appraisal of Graduation Design Excellence Award

The Oral Defense Judge Panel reviews the self recommendation plan exhibition boards at the Result Exhibition and votes on Graduation Design Excellence Award. The self recommendation plan with over half of the total votes of the Oral Defense Judge Panel will be honored with Graduation Design Excellence Award.

4. Excellent Mentor of Graduation Design

The General Mentor of topic selecting enterprise and college mentors under whose mentorship the designs are honored with Graduation Design Rank Awards or Graduation Design Excellence Award will be rated Excellent Mentor of Graduation Design of the year.

5. Best Organizer of Graduation Design

The colleges organizing the proposal survey, mid-term examination and graduation oral defense and album publication of the College-Enterprise Cooperative Graduation Design Program of the year will be awarded as Best Organizer of Graduation Design.

6. Outstanding Contributor of Graduation Design

The topic selecting enterprise and key supporting enterprises for the College-Enterprise Cooperative Graduation Design Program of the year will be named Outstanding Contributor of Graduation Design.

III. Commendations

(1) At the award ceremony, the CIID officials will present the certificates to the relevant winner of Graduation Design Rank Awards, Graduation Design Excellence Award, Best Organizer of Graduation Design and Outstanding Contributor of Graduation Design.

(2) The certificates are only valid with the seal of CIID. The certificates of Graduation Design Rank Awards and Graduation Design Excellence Award are autographed by the judges of the Oral Defense Judge Panel for commemoration.

IV. Supplementary Provisions

(1) The Rules were issued on and have come into effect as of June 8, 2013, the right of interpretation of which rest in CIID.

(2) The Rules were first revised in May 2015, revised for the second time in May 2016, and last revised in June 2017.

CIID "室内设计6+1"校企联合毕业设计
答辩、评奖、表彰工作细则（2017修订版）

本学会依据《中国建筑学会室内设计分会CIID"室内设计6+1"校企联合毕业设计章程》，制订校企联合毕业设计答辩、评奖、表彰工作细则，指导相关工作遵照执行。

一、答辩准备

（1）参加高校每校到场指导教师不超过2名，到场学生总人数限6～8名。

（2）参加高校每校安排不超过2个毕业设计答辩方案组，参加毕业设计答辩活动、"毕业设计等级奖"评选和成果展出；此外，最多可再报送2个自荐方案，不参加毕业设计答辩和"毕业设计等级奖"评选，但参加"毕业设计佳作奖"评选和成果展出。

（3）每个毕业设计答辩方案组提前准备毕业设计答辩陈述PPT等电子文档，于毕业设计答辩活动报到时提交活动组委会。

（4）答辩前每个毕业设计答辩方案、自荐方案需分别准备成果展板3张；使用学会发布统一模板编辑，展板幅面为A0加长（900mm×1800mm），分辨率不小于100dpi，使用学会发布统一模板编辑。展板电子版须于答辩前1周发送到答辩活动承办高校指定的工作邮箱；由承办高校负责汇总打印、布展等。

（5）参加高校按《CIID"室内设计6+1"校企联合毕业设计联合毕业设计书稿排版要求》编辑书稿，于毕业设计答辩活动现场提交活动组委会。

（6）毕业设计课题方向选报。CIID"室内设计6+1"校企联合毕业设计参加高校，应结合本校毕业设计教学实际，按照当届《CIID"室内设计6+1"校企联合毕业设计"主题×——总命题×"框架任务书》（以下简称《框架任务书》）设置的毕业课题方向，选报2个毕业设计答辩方案组的课题方向（单选），并于答辩前1周报送学会。

二、答辩与评奖

1. 答辩评委会构成

（1）答辩委员会由特邀评委和高校评委组成。特邀评委由命题企业、学会（含地方专业委员会）专家、活动观察员、支持企业等在内的5～7位专家担任；答辩评委会组长一般由总命题企业专家（总导师）担任。

（2）高校评委由参加高校各推选1位毕业生指导教师担任。

2. "毕业设计等级奖"评选

（1）"毕业设计等级奖"按当届《框架任务书》设置的毕业设计课题方向分别设置一、二、三等奖，奖项设置相应比例一般为1:2:4，奖项评审可以空缺。

（2）第一轮评选。参加高校每个毕业设计答辩方案组按选报课题方向进行答辩，每组答辩陈述时间不超过15分钟，问答不超过15分钟。由特邀评委、高校评委共同填写选票，进行排序评选（例如，"1"为建议排序第一，"2"为建议排序第二等，依次类推）。活动组委会负责排序选票统计，形成相应设计专题"毕业设计等级奖"评选建议排序。

（3）第二轮评选。高校评委须回避，由特邀评委以"毕业设计等级奖"评选建议排序为基础，对照答辩方案进行审议评选，确定当届"毕业设计等级奖"获奖方案。

3. "毕业设计佳作奖"评选

答辩评委会对参加成果展示的高校自荐方案展板进行评议，对是否认定为"毕业设计佳作奖"方案进行投票；同意票数超过答辩评委会总人数1/2（含）的高校自荐方案，确定获得佳作奖。

4. 优秀毕业设计导师

获得"毕业设计等级奖""毕业设计佳作奖"作品的命题企业总导师、参加高校导师成为当届"优秀毕业设计导师"。

5. 毕业设计最佳组织单位

承办当届校企联合毕业设计活动开题踏勘、中期检查、毕业答辩、总辑出版的高校成为"毕业设计最佳组织单位"。

6. 毕业设计突出贡献单位

负责当届校企联合毕业设计活动的命题企业、重要支持企业等成为"毕业设计突出贡献单位"。

三、表彰奖励

（1）在活动颁奖典礼上，由学会领导分别向毕业设计等级奖、毕业设计佳作奖、优秀毕业设计导师、毕业设计最佳组织单位、毕业设计突出贡献单位的获得者颁发证书。

（2）奖励证书由学会盖章有效。毕业设计等级奖、优秀毕业佳作奖的证书印有活动答辩评委会评委签名，以示纪念。

四、附则

（1）本规则2013年6月8日公布施行，由中国建筑学会室内设计分会负责解释。

（2）本规则2015年5月第一次修订，2016年5月第二次修订，2017年6月第二次修订。

調研踏勘
Investigation and Survey

CID『室内设计 6+1』2017（第五届）
校企联合毕业设计
CID"Interior Design 6+1"2017(Fifth Session)
University-Enterprise Cooperative Graduation Design

国匠承启卷——传统民居保护性利用设计
Craftsmanship Heritage
——Design for the Protective Utilization of Traditional Folk Houses

貳

國匠承啓卷——傳統民居保護性利用設計

同济大学
Tongji University

CIID "室内设计 6+1" 2017（第五届）校企联合毕业设计
CIID "Interior Design 6+1" 2017(Fifth Session)University-Enterprise Cooperation Graduation Design

上海近代黑石公寓周边环境保护与更新设计

小组成员：

常馨之 Chang Xinzhi　　　李淑一 Li Shuyi　　　安麟奎 An Linkui
张雨缇 Zhang Yuti　　　　周雨桐 Zhou Yutong　　谢丹妮 Xie Danni

基地概况 ‖ Site Profile

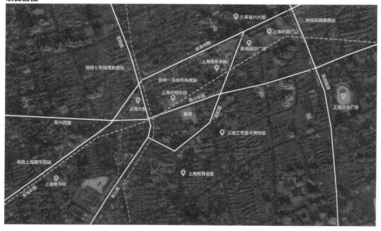

　　黑石公寓或者复兴公寓，是中国第一个专门为居住在上海的外籍人士设计的豪华公寓。这栋六层建筑建于1924年，通过从英国进口的水泥和石头建造出灰暗的外观效果。

　　建筑位于上海徐汇区复兴中路（原拉斐德路）中段，毗邻街角的汾阳路。它位于上海旧法租界地区的中心地段。它同样也是上海市历史保护建筑。

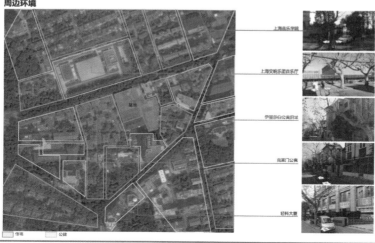

調研踏勘

现状建筑

现状空间
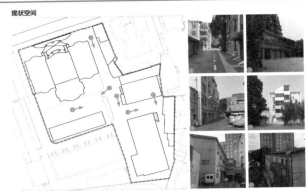

现状建筑

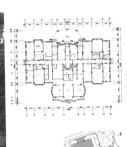

现状建筑

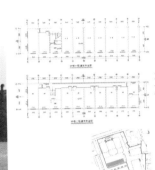

现状建筑

现状建筑

基地道路交通分析

基地周边500m范围内有两处地铁站及多个公交站点，具有很强的可到达性。

基地与道路关系图
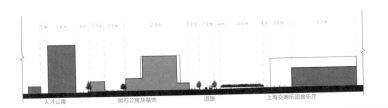

黑石公寓与上海市交响乐音乐厅分别位于复兴中路的两侧，这是一种关于古典力量和现代力量的对比。

附件：建筑与场地图

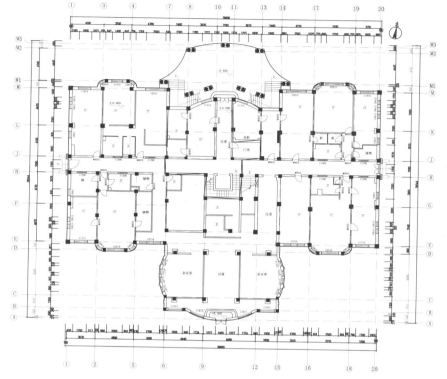

黑石公寓首层平面图

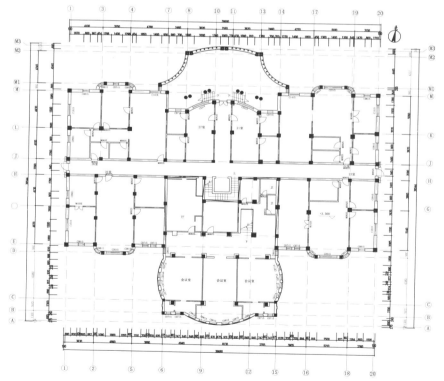

黑石公寓二层平面图

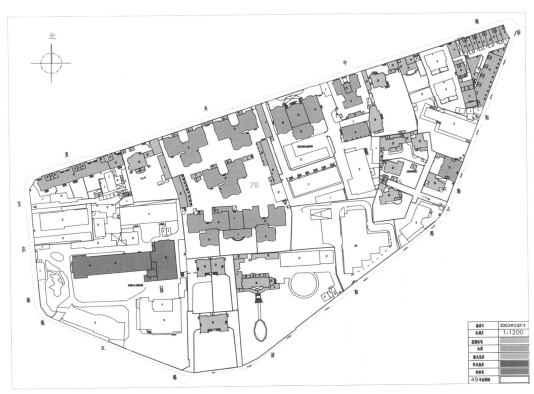

基地总平面图

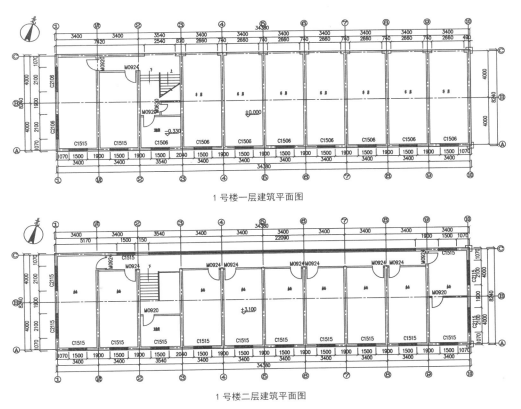

1号楼一层建筑平面图

1号楼二层建筑平面图

1号楼平面图

华南理工大学
South China University Of Technology

CIID"室内设计 6+1"2017（第五届）校企联合毕业设计
CIID"Interior Design 6+1"2017(Fifth Session)University-Enterprise Cooperation Graduation Design

泮塘五约村民居更新设计

小组成员：

郭嘉敏 Guo Jiamin　　　　谈　志 Tan Zhi

基地概况 ‖ Site Profile

泮塘，曾经被时代遗忘，现重新被提起。我们看到的它的衰落，人口的老龄化，政府机关、开发商与原住民的困斗等问题，其背后有极其复杂的原因。

地区经济体的流失直接导致地区的衰败。衰败则加重经济体的流失，如此恶性循环，这是诱因一。

人口结构走向老龄化，经济同时衰落无法引入新的人口，老龄化问题走向无解，这是诱因二。

我们设想，我们需要引入新的地区使用者，他们是新的经济个体。我们要做合适的设计，能使新人群与老人群融洽共处，甚至使老人们融入新的经济体，探索村落的记忆，重新激活中心区域，延续这里的故事。

國匠承啟卷——傳統民居保護性利用設計

区位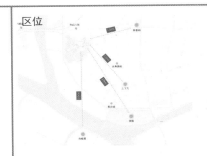

交通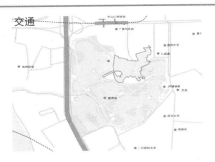

功能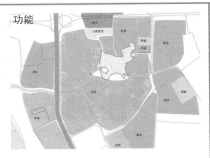

泮塘地区位于广州旧城区，临近荔湾湖，区位、交通条件优越，与周边城市重点区域联系便捷。

泮塘五约地处荔湾区中心，被荔湾湖环绕，景色优美，北侧是区域交通中心，周边有多处历史文化遗产。

泮塘五约周边以住宅和公园为主，住宅包括保留村落和小区住宅以及相应的配套设施。

项目调研 ‖ Project Research

视频片段截图

泮塘五约直街

交流　自给自足　邻里

散点性

周期性

自发性

多样性

國匠承啓卷——傳統民居保護性利用設計

问题研究 || Problem Research

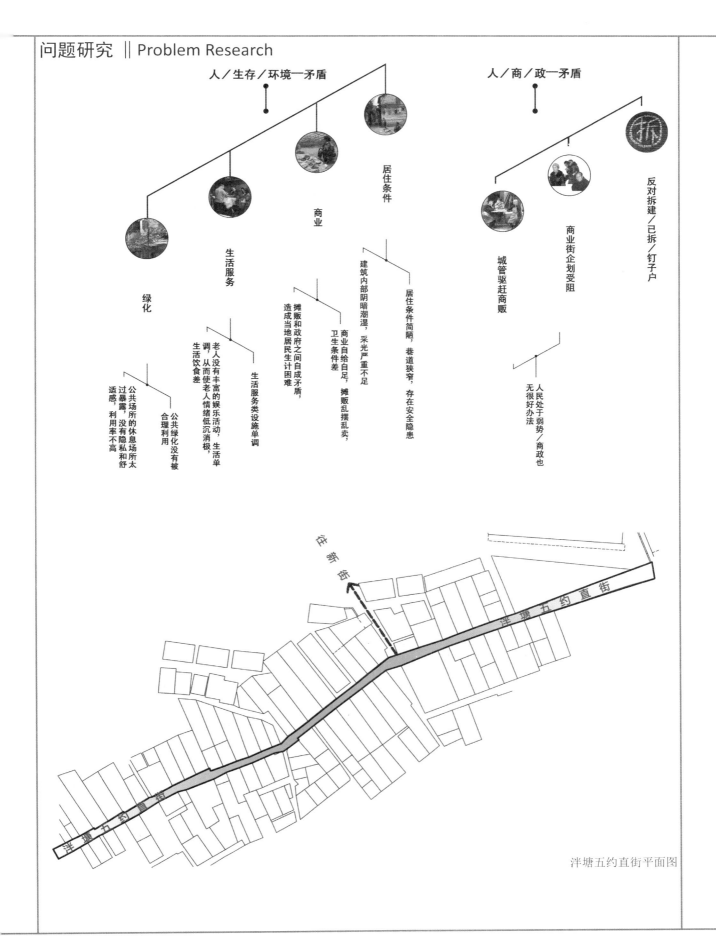

泮塘五约直街平面图

华南理工大学
South China University of Technology

CIID "室内设计 6+1" 2017（第五届）校企联合毕业设计
CIID "Interior Design 6+1" 2017(Fifth Session) University-Enterprise Cooperation Graduation Design

广东省广州市珠光路侨商街地块民居更新设计

小组成员：

林嘉辉　Lin Jiahui　　　　张裕麟　Zhang Yulin　　　　卢惠兴　Lu Huixing
陈小帆　Chen Xiaofan

基地概况 ‖ Site Profile

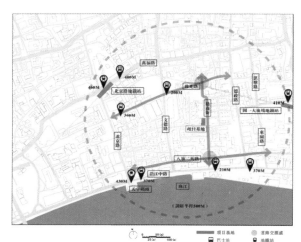

建筑位于珠光路小区内，有侨商街穿过，南北端分别与珠光路与八旗二马路相交。小区周围有多个公交车站、地铁站，航运有天字码头，交通种类丰富，区位优势明显。

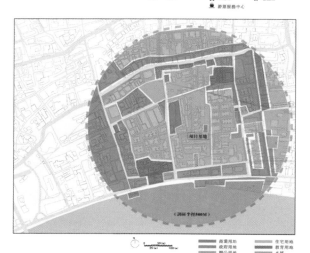

小区周边的业态丰富，用地类型众多，周边娱乐和基础设施配套较为完善，包括餐饮、购物商店和大型停车场等。其中住宅用地和商业用地占比很大，商业用地中有很大一部分都是小商品买卖、周边还有该种类大型的批发市场以及电商团体。另外是办公的区域，但不是特别多，比办公用地更少的是绿化及活动用地。

建筑分析

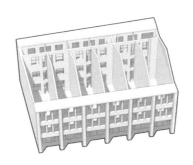

　　侨商街洋房为三层高的联排建筑物，是广州少有的南洋住宅、建筑风格是广州传统骑楼形式，以前被居民用来商住两用。建筑的内部空间结构陈旧，如高窄的楼梯、狭长的房型，已不能满足现代人们的居住需求。

竖向空间联系

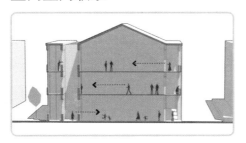

　　原有空间仅靠一条直通的狭长楼梯进行楼层之间的联通，缺少空间之间的交流与联系。

景观植被

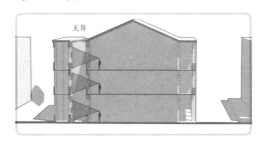

　　建筑后部有绿色景观，但是视线被建筑后部的副楼挡住，而室内又缺少绿植。

采光通风

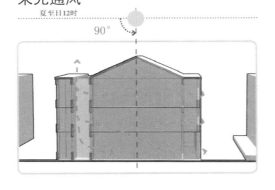

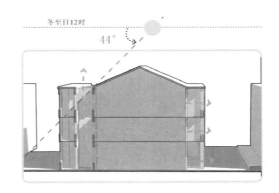

　　原先室内空间进深大，狭窄成条，很大程度上依赖天井进行采光通风，因不能满足整栋建筑需求而导致室内空间尤其是底部空间潮湿阴暗。

受众分析

　　小区的主要人群为30～50岁的成年人，但主要活动人群却是50岁以上的老年人。

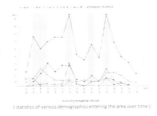

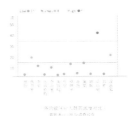

附件：建筑与场地图

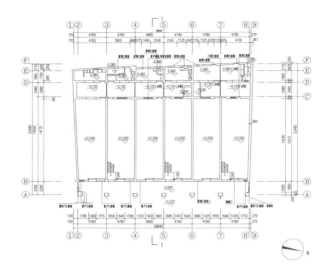

首层平面图

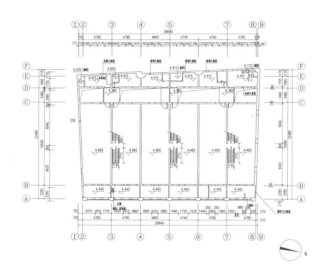

二层平面图

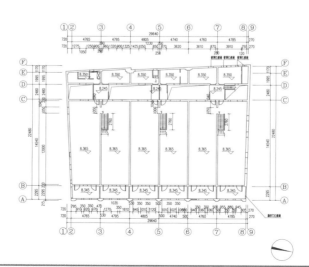

三层平面图

調研踏勘

立面图

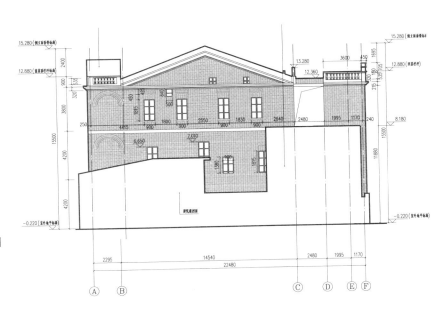

侧立面图

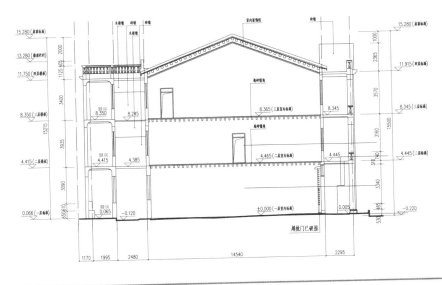

剖立面图

哈尔滨工业大学
Harbin Institute of Technology

CIID"室内设计 6+1"2017（第五届）校企联合毕业设计
CIID"Interior Design 6+1"2017(Fifth Session)University-Enterprise Cooperation Graduation Design

哈尔滨道外传统里院空间改造更新设计

小组成员：

于　璐 Yu Lu	李　焱 Li Yan	胡一非 Hu Yifei
单　昳 Shan Yi	樊逸冰 Fan Yibing	郭冰燕 Guo Bingyan
刘名淞 Liu Mingsong	曹在笑 Cao Zaixiao	李佳星 Li Jiaxing
刘金哲 Liu Jinzhe	李冬霓 Li Dongni	程子芙 Cheng Zifu
孙梦扬 Sun Mengyang		

基地概况 ‖ Site Profile

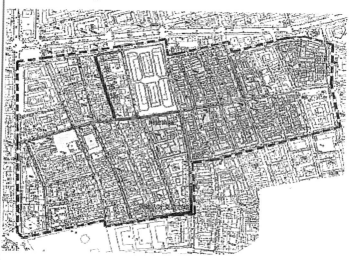

道外区旧称傅家甸，地处哈尔滨市区中东部，北临松花江，作为哈尔滨市的开埠地，1956年改称道外区。道外区是哈尔滨这座历史文化名城历史最悠久的重要组成部分，是关道文化和"闯关东"文化的汇集地，以"中华巴洛克"片区为主体的道外历史街区是重要的建筑文化遗产。道外传统里院是旧城区中极具特色的居住和商业空间。

核心区公共建筑分布现状：许多建筑如今作为公用，服务于办公、商业、图书展览、影院、医院、公共厕所、餐饮等。

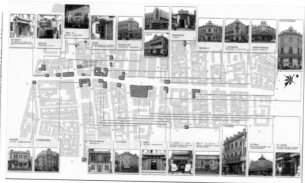

核心区老字号分布现状：在此区域内，仍屹立着众多老字号。古色古香，韵味十足。

建筑风格 ‖ Style of Architecture

作为特定历史时期、特定社会背景条件下产生的风格类型，中华巴洛克建筑可以说是哈尔滨近代建筑的宝贵财富。

在今天的哈尔滨，我们依然能看到建筑的样式保留了建城初期的哈尔滨印象，政府将这种建筑风格传下来，保有了哈尔滨的特色，当代建筑中不论居民建筑还是商业建筑中，都依然保有这种折衷风格，所以中华巴洛克建筑不论是从过去的存在价值还是如今的历史价值来看，都具备非常重要的意义。

女儿墙

窗

线脚与装饰

建筑空间特点 ‖ Feature of Space

面对外来文化的影响，早期的中国工匠基于自己的理解创造出了中华巴洛克历史街区形态多样的建筑。平面构成方面北方里院民居的特征被保留下来；沿街立面上则集中了模仿西式风格的装饰元素；垂直流线围绕里院中庭进行组织。

社区服务中心位于小区中心，其余建筑均为居住功能。

平面构成　　沿街立面　　垂直流线

空间形态 ‖ Form of Space

复杂的院落肌理可以归纳为四种基本形态

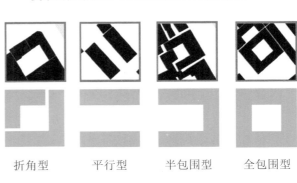

折角型　　平行型　　半包围型　　全包围型

西安建筑科技大学
Xi'an University of Architecture and Technology

CIID"室内设计 6+1"2017（第五届）校企联合毕业设计
CIID"Interior Design 6+1"2017(Fifth Session)University-Enterprise Cooperation Graduation Design

思洞·漫步——榆林赵庄窑洞保护性再利用

小组成员：

左怀腾 Zuo Huaiteng　　杨　月 Yang Yue　　田泽宇 Tian Zeyu
徐瑞显 Xu Ruixian　　　夏兆康 Xia Zhaokang　史文媛 Shi Wenyuan

基地概况 ‖ Site Profile

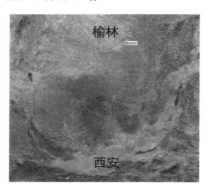

赵庄村位于陕北榆林城区东南 10km 处，三鱼公路穿境而过，毗邻 210 国道，交通便利。全村辖贾家村，南山沟等 6 个自然村组成，村域面积 6.87km²，总人口 2050 人，是古塔镇最大的一个村子。

基地具体范围

調研踏勘

建筑组团

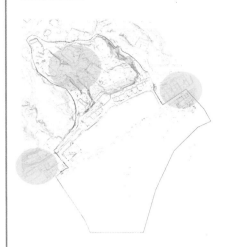

基地建筑组团形式有居住区、废弃建筑区、待修复建筑区。

地形特征

地形主要分为平地区和山地区，属于典型的陕北黄土高原特征。

流线组织

穿行基地分为车行和人行两条道路。

建筑性质

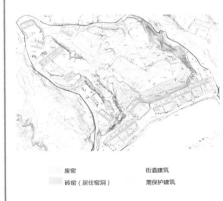

废窑　　　街道建筑
砖窑（居住窑洞）　　需保护建筑

基地建筑废窑已不满足于生活需求大多为生土窑；街道建筑为新建窑洞；砖窑现在仍有居民居住；需要保护的建筑年久失修但结构保留完好。

建筑与环境关系

与环境相协调的建筑　　新建建筑
与环境不协调的建筑

简易砖房建筑既不能满足村民对生活的追求，又与周边环境不协调，大部分废弃掉。

建筑年代

清代　　　2000年后

张家大院建筑有300余年，有较高的保护价值；生土窑洞有50～100年，大多已经废弃；砖石窑洞小于50年，现在仍有老人居住；新型砖房建筑为现代建造。

用地布局

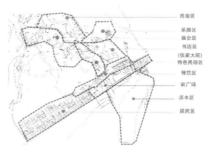

赵庄整体用地主要分为种植区、乡村旅游区、居民区、自然区。

功能结构

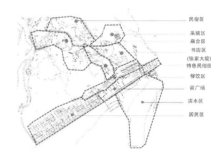

经过初步规划下的功能区。

道路系统

总体的交通路线和景点浏览路线。

窑洞空间模型和照片

- 要留住建筑原有的特色和传统美。
- 要对原来窑洞室内的优缺点进行分析。
- 要对传统窑洞建筑细节保留和创新。
- 要对传统结构有所传承。
- 要对当地的资源充分利用合理利用。

建筑立面和建筑元素

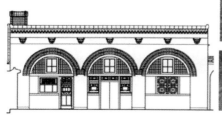

窑洞细节结构

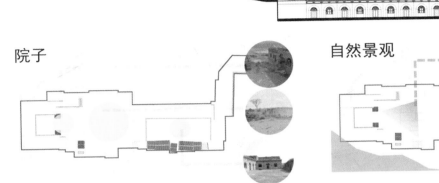

院子

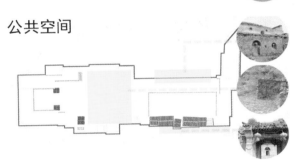

自然景观

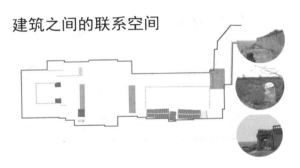

公共空间

建筑之间的联系空间

- 原建筑的残缺需要挑选保护
- 原来院子无空间秩序需要整体规划交通和空间
- 原只有两个独立残破的院子现要将两院联系起来
- 周边的自然风景要和院子结合起来
- 要有现代的生活元素和传统元素和建筑元素相结合

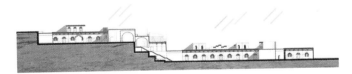

没有加建高度过高的建筑，建筑根据地形递进。

中间依地形有加建可能性，两边院落保留原貌并加以保护利用。

附件：建筑与场地图

赵庄场地地形图

赵庄场地平面图

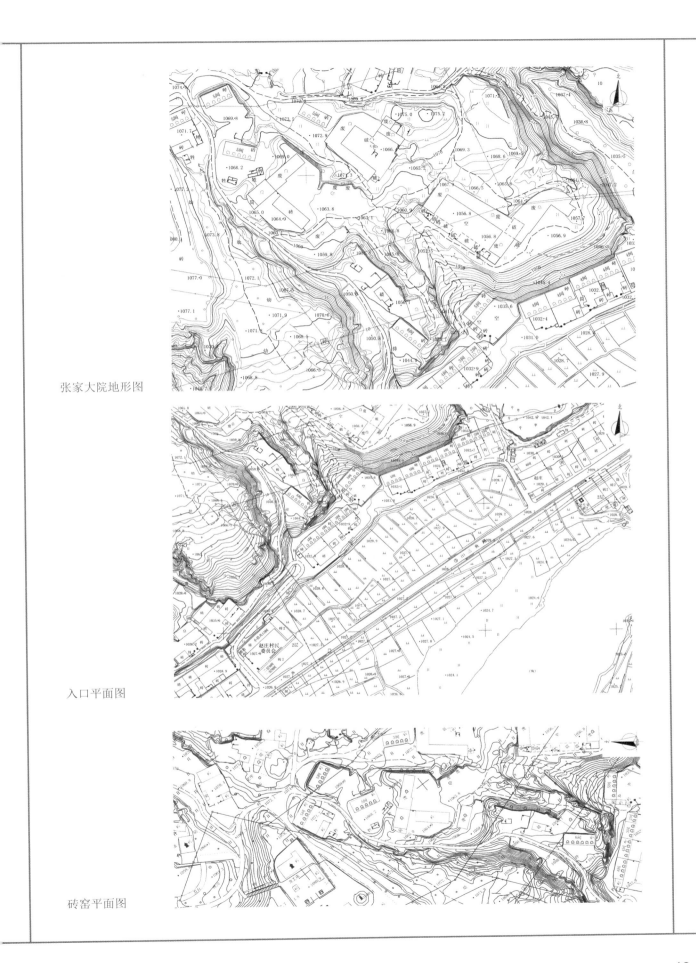

张家大院地形图

入口平面图

砖窑平面图

西安建筑科技大学
Xi'an University of Architecture and Technology

CIID"室内设计 6+1"2017（第五届）校企联合毕业设计
CIID"Interior Design 6+1"2017(Fifth Session)University-Enterprise Cooperation Graduation Design

情系黄土　重返家园——地坑院保护性利用设计

小组成员：

季　然 Ji Ran　　　　赵　涛 Zhao Tao　　　　崔为天 Cui Weitian

基地概况 ‖ Site Profile

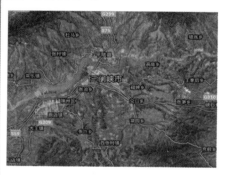

　　陕州即今三门峡市陕县，东据崤山关连中原腹地，西接潼关、秦川扼东西交通之要道，陕州北营村现存32孔窑洞，村民约80余人，村落周围农田环绕，村落的平均海拔约700m。

基地具体范围

基地建筑组团形式有居住区、废弃建筑区和待修复建筑区。

地形主要分为平地区和山地区，属于典型的陕北黄土高原特征。

穿行基地分为车行和人行两条道路。

优点：
- 工程造价低廉且环保。
- 构建方式独特。

缺点：
- 基地交通不便。
- 建筑内部采光通风差。
- 建筑地上地下联系不紧密。

基地剖立面图

附件：建筑与场地图

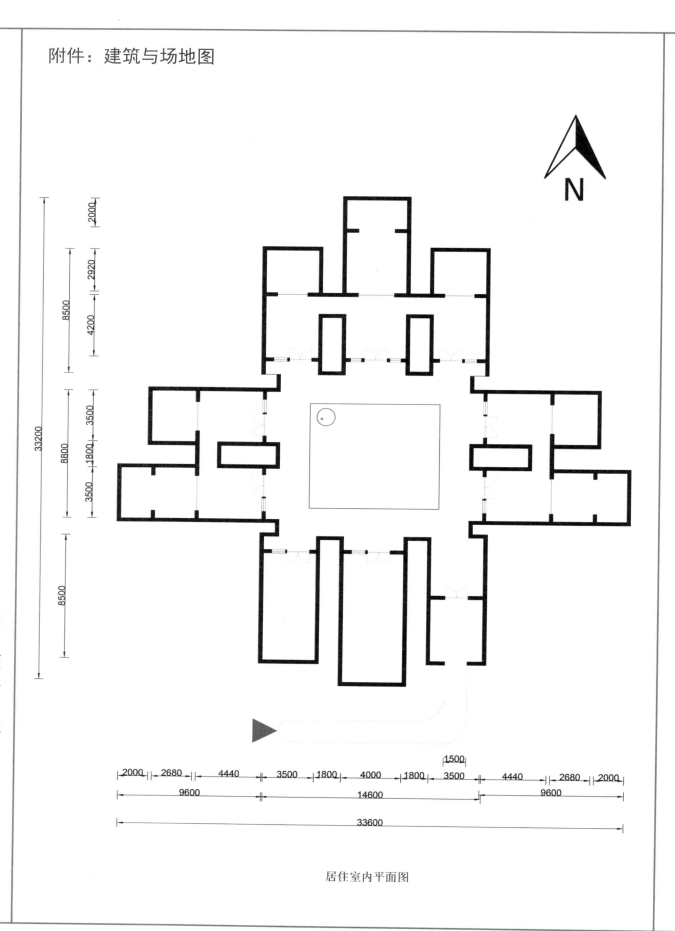

居住室内平面图

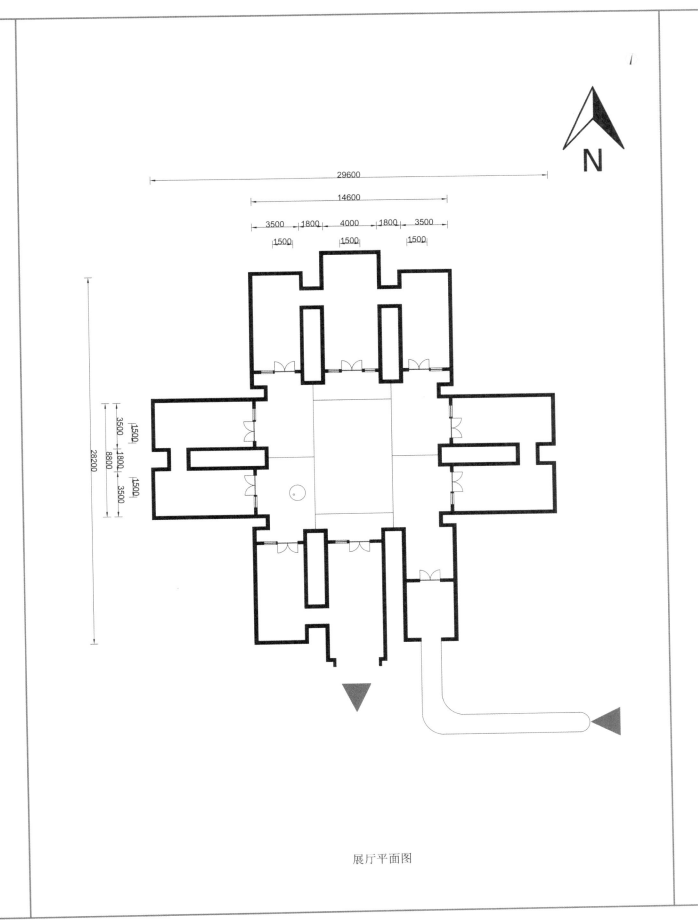

展厅平面图

北京建筑大学
Beijing University of Civil Engineering and Architecture

CIID"室内设计 6+1"2017（第五届）校企联合毕业设计
CIID"Interior Design 6+1"2017(Fifth Session)University-Enterprise Cooperation Graduation Design

社区文化活动中心

小组成员：

汤博文 Tang Bowen　　　　蔡明秀 Cai Mingxiu　　　　刘佳蕊 Liu Jiarui

基地概况 ∥ Site Profile

"家巧儿"社区活动中心位于北京市东城区北二环内青龙胡同西北角北邻雍和大厦等高层办公楼，南邻胡同居民房，西邻雍和宫柏林寺。涉及范围约 24 万 m²。活动中心场地面积约 4800 m²，包含传统四合院一栋，戏楼一座，现代建筑一间，建筑面积共 1240 m²。

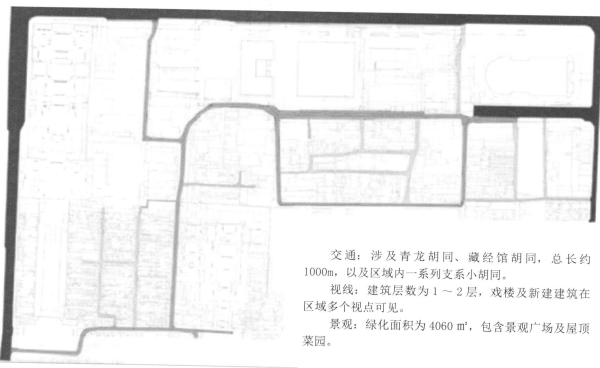

交通：涉及青龙胡同、藏经馆胡同，总长约 1000m，以及区域内一系列支系小胡同。

视线：建筑层数为 1～2 层，戏楼及新建建筑在区域多个视点可见。

景观：绿化面积为 4060 m²，包含景观广场及屋顶菜园。

健康的生活方式让我们不用逃离都市，更勇敢和有创造性地面对生活。

有生命力的东西总有成长过程，所以不用急，慢慢来。

共生系统中，菜地为餐厅提供新鲜蔬菜，餐厅的果皮和咖啡渣又作为堆肥回馈菜地。

鲜嫩诱人的青菜苗，色泽艳丽的小西红柿，青翠欲滴的小黄瓜……

在种植的过程中，原本陌生的邻居在一起施肥浇水，分享种菜心得，不但种出了各色瓜果蔬菜，也种出了邻里间的感情，这是最大的收获。

透过耕作，增加人与人、人与食物以及人与土地之间的互动。

"城市农夫"运动身体、舒缓压力，新鲜瓜果蔬菜，缩短农场到餐桌的距离。水泥森林回归自然。

即摘即食。

- 种植土
- 排水层
- 过滤层
- 保护层
- 根阻防水层
- 找平层

750mm种植土
120mm蛭石排水层
过滤层：250~300g/m² 的聚酯无纺布
20mm混凝土保护层
3mmAPP聚酯卷材和3mm抗根卷材做防水
20厚1：3水泥砂浆找平层
原建筑结构层

色彩分析 ‖ Color Analysis

春

2月初播种，4月呈现最佳状态，品种搭配90%按照设计进行。

色彩搭配和谐，病虫害较少。

可采摘豌豆、西红柿、甜椒辣椒、西葫芦、莴笋、菠菜、香椿。

可种植黄瓜、油麦菜、马铃薯、洋葱。

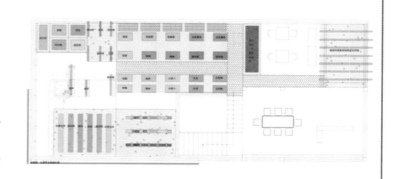

夏

夏季，蔬菜疯长，密度比较拥挤，高度遮挡视线，局部观赏效果不佳。

周边草坪被陆续开辟用做露天育苗实验基地。

可采摘冬瓜、茄子、西红柿、西蓝花、玉米。

可种植土豆、甘蓝、萝卜、白薯、绿豆、蚕豆。

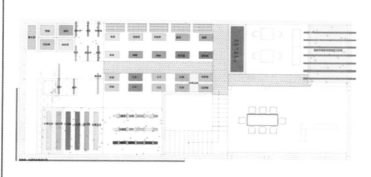

秋

秋天是收获的季节，也是分享的季节。果实类蔬菜占比60%左右，从景观效果而言，大批蔬菜收掉后清场整地补种冬季菜苗，高低不齐，景观效果不佳。

可采摘黄瓜、扁豆、南瓜、葫芦、茄子、紫薯。

可种植茄子、西红柿、芸豆、辣椒。

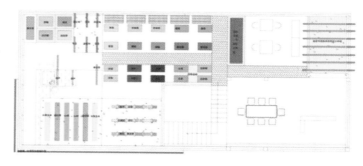

冬

冬季品种几乎全部替换为可露天越冬的蔬菜品种，种类较少且生长缓慢，观赏植物数量减少。冬季病虫害较少。

可采摘白菜、萝卜、芥蓝、姜。
可种植黄瓜、油麦菜、花菜、冬瓜。

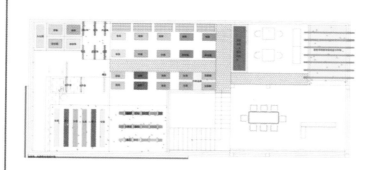

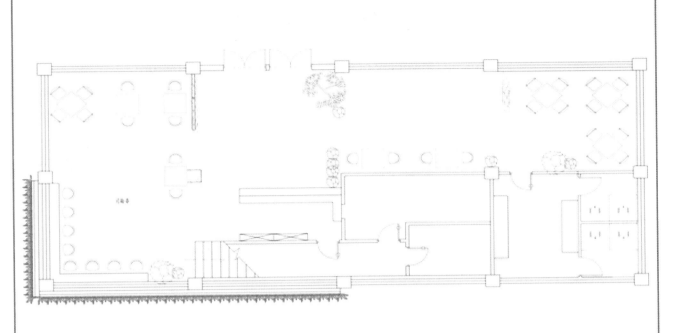

餐厅一层平面图

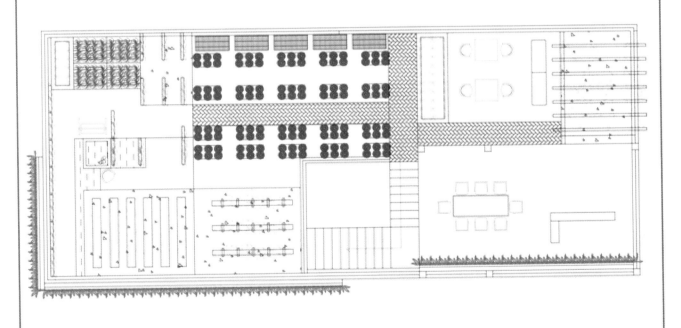

餐厅二层平面图

北京建筑大学
Beijing University of Civil Engineering and Architecture

CIID"室内设计 6+1"2017（第五届）校企联合毕业设计
CIID"Interior Design 6+1"2017(Fifth Session)University-Enterprise Cooperation Graduation Design

青龙胡同四合院与戏楼改造设计

小组成员：

尤　昀 You Yun　　　　孙卫圣 Sun Weisheng　　　　张乐情 Zhang Leqing

基地概况 ‖ Site Profile

北京城的肌理在一百多年前就初步形成，分布在东西两侧的王公贵族与富商名流，经过世代相传，其居住场所——四合院已经颇具规模与格局。散落在京城各处寻常百姓的院子也和胡同完成了融洽的结合。

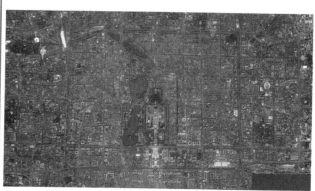

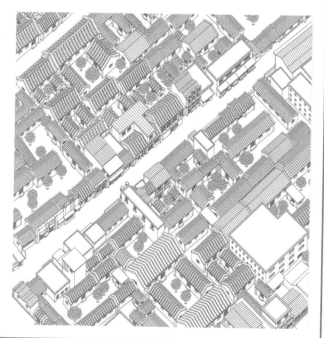

國匠承啓卷——傳統民居保護性利用設計

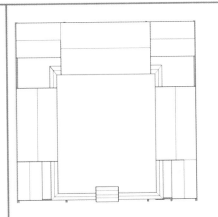
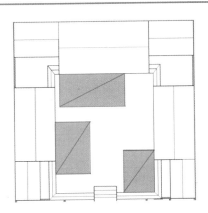
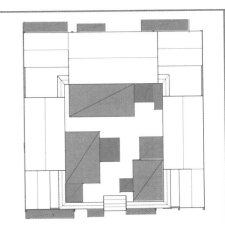

以一间四合院为基本模型，其家族第一代建造并生活于此，至第二代，因居住所需，在庭院内搭建房屋；至第三代，继续在四合院外围搭建。

- 调研范围总面积：共178平方米。
- 调研胡同：藏经馆胡同、青龙胡同、炮局头条、炮局二条、炮局三条、炮局四条、炮局胡同、育树胡同、育树二条、育树四条、戏楼胡同、戏楼二条。共13条胡同。
- 调研院落：共96院。
- 调研民居：共17户。

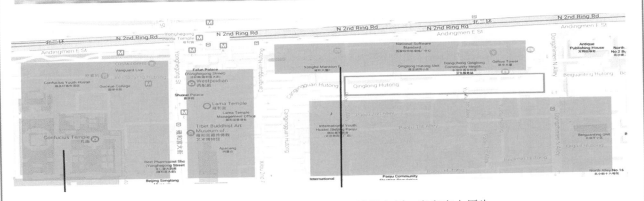

五道营、国子监、方家胡同等新兴文化产业区及文化创意园区，聚集了大量青年在此休闲、生活、办公。

歌华大厦、雍和宫大厦为文创公司的聚集地，大量上班族日常在此出没。

附件：建筑与场地图

四合院一层平面图

調研踏勘

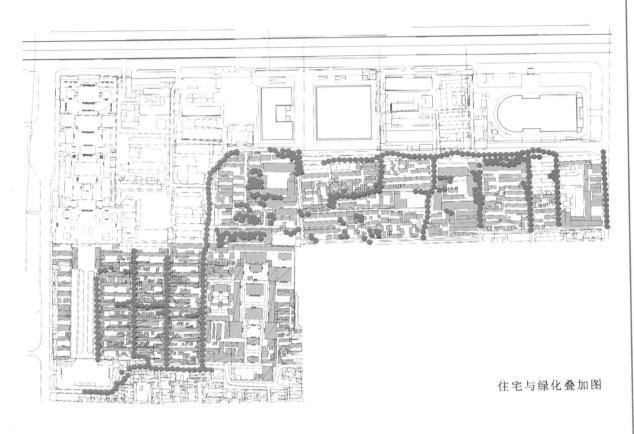

住宅与绿化叠加图

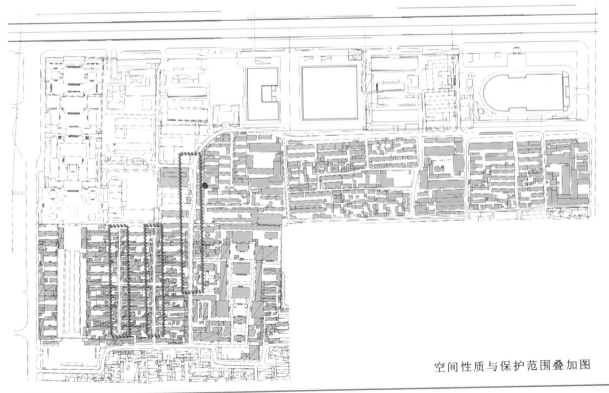

空间性质与保护范围叠加图

南京艺术学院
Nanjing University of the Arts

CIID"室内设计 6+1"2017（第五届）校企联合毕业设计
CIID"Interior Design 6+1"2017(Fifth Session)University-Enterprise Cooperation Graduation Design

甘熙宅第保护性利用设计

小组成员：

林秋霞 Lin Qiuxia	王依丽 Wang Yili	章伶钰 Zhang Lingyu
朱　彦 Zhu Yan	赵雨琦 Zhao Yuqi	黄雅君 Huang Yajun

基地概况 ‖ Site Profile

甘熙宅第位于南京城市中心，地处南京升州路与中山南路交界地段，与熙南里历史商业街区以及南捕厅历史文化街区相邻，又靠近新街口以及夫子庙商圈，总面积2.1万余平方米，是南京现有面积最大，保存最完整的私人民宅。

甘熙宅第现已开辟为南京民俗博物馆，2006年被列为全国重点文物保护单位。

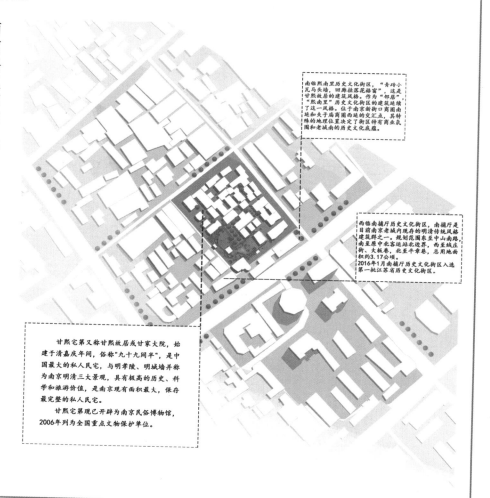

南临熙南里历史文化街区，"青砖小瓦马头墙，回廊挂落花格窗"，这是甘熙故居的建筑风格，作为"邻居""熙南里"历史文化街区的建筑延续了这一风格。位于南京新街口商圈南延和夫子庙商圈西北的交汇点，其特殊的地理位置决定了街区特有的商业氛围和老城南的历史文化底蕴。

西临南捕厅历史文化街区，南捕厅是目前南京老城区内现存的明清传统风格建筑群之一。规划范围东至中山南路，南至原中北客运站北边界，西至绒庄街、大板巷，北至平章巷，总用地面积约3.17公顷。2016年1月南捕厅历史文化街区入选第一批江苏省历史文化街区。

甘熙宅第又称甘熙故居或甘家大院，始建于清嘉庆年间，俗称"九十九间半"，是中国最大的私人民宅，与明孝陵、明城墙并称为南京明清三大景观，具有极高的历史、科学和旅游价值，是南京现有面积最大，保存最完整的私人民宅。
甘熙宅第现已开辟为南京民俗博物馆，2006年列为全国重点文物保护单位。

周边环境 || Surrounding Environment

甘家大院周边商区发展成熟，配套设施良好，吃、购、住、行都十分便利，周边公交站地铁站数量众多，道路四通八达，为来访者提供了很好的出行条件。

南临熙南里文化街区

西临南捕厅旧址

北临私人工作室

东临商业区

建筑分析 || Analysis of Architecture

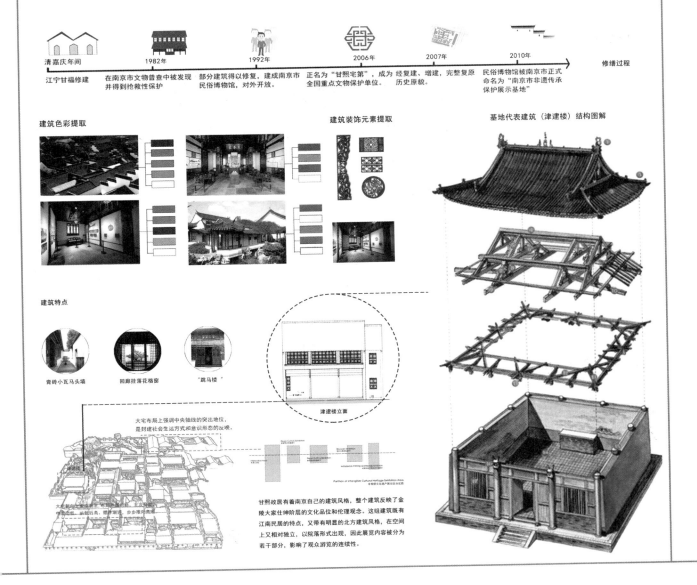

建筑控制要求 || Construction Control Requirements

- 高度控制：新建建筑高度控制在 1~2 层，檐口高度不超过 7 m。
- 整体风貌控制：建筑色彩应以黑、白、灰为主色调。
- 新建建筑建设控制：采用与整体风貌相协调的手法，严格控制。
- 建筑高度与建筑形式，确保新建建筑与周围环境相协调。

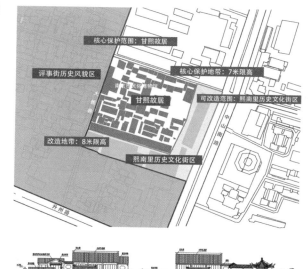

甘家大院建筑高度为 1~2 层

历史文化 || History and Culture Requirements

甘家大院最初称为"友恭堂"。明末，一支甘氏族人离开丹阳甘村，进入城内，开始以务农、经营田产为生，至清乾嘉之际，南京丝织业大盛，甘国栋（邑士）率子行商，经营"剪绒、江绸、贡缎、棉纱、布帛"，家境逐渐殷实。嘉庆初年，甘国栋在时称府西大街的南捕厅买下一块宅基地，开始营建房屋，嘉庆己未（1799年）正式迁居于此，并取堂名曰"友恭堂"。

甘国栋创立"友恭堂"

来访人群分析 || Visiting Groups Analysis

SWOT 分析 || SWOT Analysis

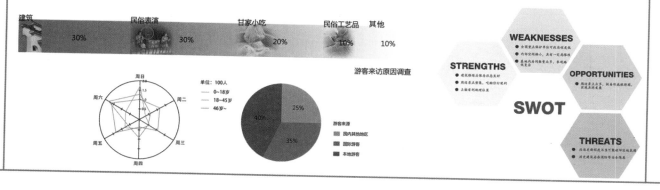

资源优势 ‖ Resource Advantage

甘家大院为全国重点保护单位，建筑本身独具特色。

基地花园提供了活动场地。

天井内设人物场景雕塑。

展览多处设有互动体验区增加了展览的趣味性。

现存问题分析 ‖ Analysis of Existing Problems

房间开口较多，内部行走路线混乱。

古建自身条件远达不到现代展陈要求，古建密封不严导致温湿度调节等问题难解决。

房间空间矮小，大型展览无法体现。

展览电器设备等外露，影响美观。

前厅空间狭小，功能配置不足。

设计要求 ‖ Design Requirements

保护：
街区是目前南京老城内现存的明清传统风格建筑群之一，其历史格局清晰、传统风貌完整、历史遗存丰富，具有较高的历史价值、文化价值、景观价值和旅游价值。

利用古建形制
处理好六大空间关系，将展览融入古建中，做到馆园合一

园内 馆内 柜内
园外 馆外 柜外

保护 PROTECR
空间 SPACE
文化 CULTURE
信息 INFORMATION
改造 REFORM
利用 EXPLOIT
传承 INHERIT

本次设计为甘熙故居及周边熙南里历史商业街区一起综合考虑重新定位和更新改造，可根据市场需求自定业态类型进行设计工作，设计任务包括：前期调研及业态可行性分析、周边环境总体规划、单体建筑设计、室内利用设计等。

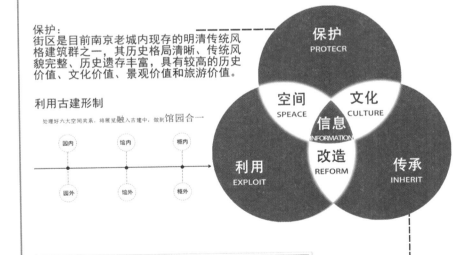

传承建筑本身及其历史文化与价值

浙江工业大学
Zhejiang University of Technology

CIID "室内设计 6+1" 2017（第五届）校企联合毕业设计
CIID "Interior Design 6+1" 2017(Fifth Session) University-Enterprise Cooperation Graduation Design

浙江省台州市黄岩富山乡半山村地块民居更新设计

小组成员：

高　煊 Gao Xuan	陈玫宏 Chen Meihong	胡安琪 Hu Anqi
周　睿 Zhou Rui	洪朝艳 Hong Chaoyan	杨　洁 Yang Jie

基地概况 ‖ Site Profile

半山村始建于北宋年间，距今已有近 900 年历史。村内有条建于明清时期的黄永古道，是古代台州、温州之间的重要商旅驿站。村庄海拔 491m，人口 576 人，面积 $200hm^2$（90% 是林地和耕地）。南部竹海环绕，长决线穿村而过。

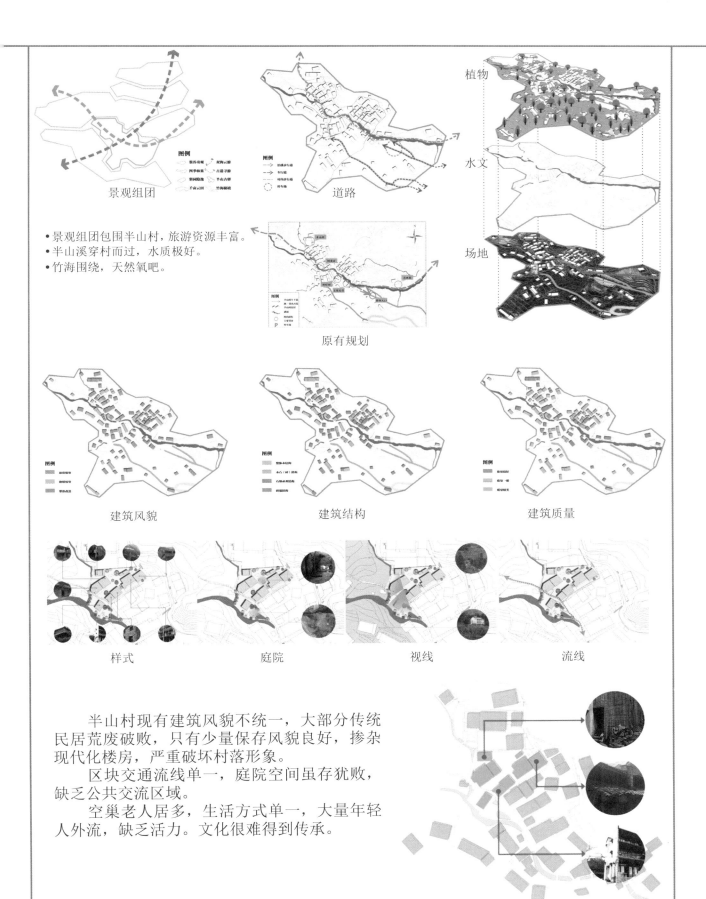

中期檢查
Medium-term Inspection

CIID「室内设计 6+1」2017（第五届）
校企联合毕业设计
CIID"Interior Design 6+1"2017(Fifth Session)
University-Enterprise Cooperative Graduation Design

国匠承启卷
——传统民居保护性利用设计
Craftsmanship Heritage
——Design for the Protective Utilization of Traditional Folk Houses

參

國匠承啓卷——傳統民居保護性利用設計

同济大学
Tongji University

CIID"室内设计 6+1"2017(第五届)校企联合毕业设计
CIID"Interior Design 6+1"2017(Fifth Session)University-Enterprise Cooperative Graduation Design

上海近代黑石公寓及其周边环境保护与更新设计

小组成员：

张雨缇 Zhang Yuti　　　周雨桐 Zhou Yutong　　　谢丹妮 Xie Danni

街区调研 ‖ Block Research

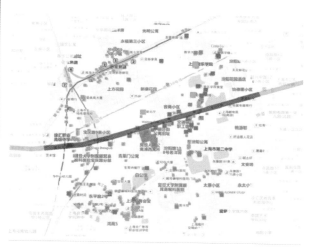

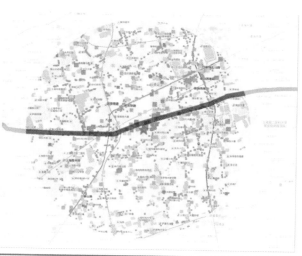

基地周边1KM业态数量

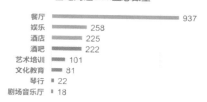

基地周边500M业态数量

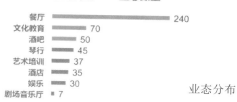

业态分布

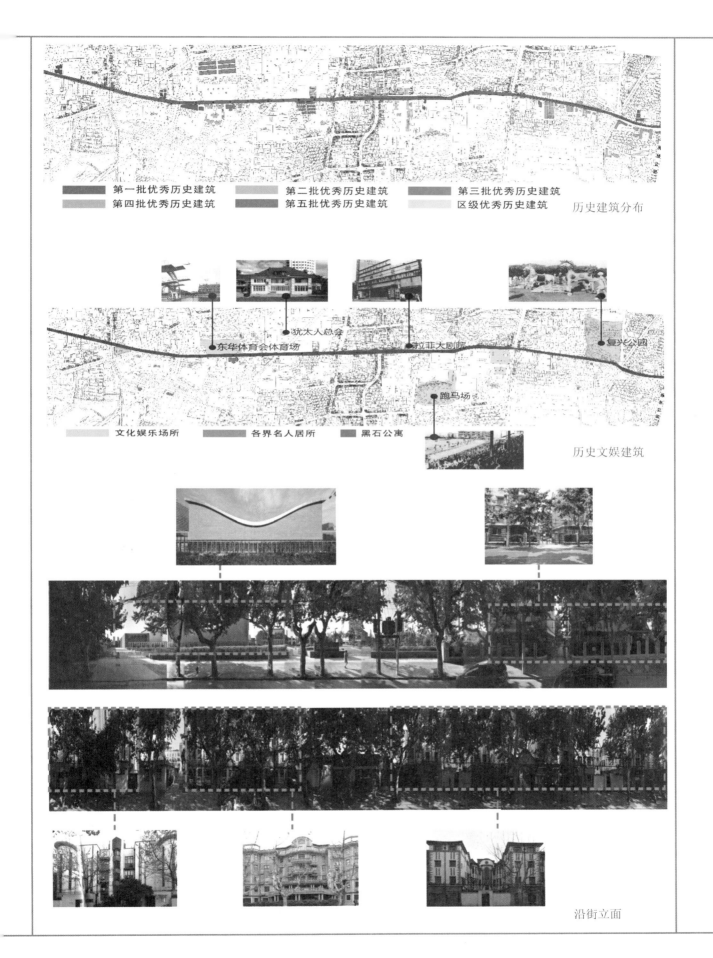

中期檢查

历史建筑分布
- 第一批优秀历史建筑
- 第二批优秀历史建筑
- 第三批优秀历史建筑
- 第四批优秀历史建筑
- 第五批优秀历史建筑
- 区级优秀历史建筑

历史文娱建筑
- 文化娱乐场所
- 各界名人居所
- 黑石公寓

东华体育会体育场　犹太人总会　寒拉菲大剧院　复兴公园　跑马场

沿街立面

同济大学
Tongji University

CIID"室内设计 6+1"2017（第五届）校企联合毕业设计
CIID"Interior Design 6+1"2017(Fifth Session)University-Enterprise Cooperative Graduation Design

上海近代黑石公寓及其周边环境保护与更新设计

小组成员：

李淑一 Li Shuyi　　　　常馨之 Chang Xinzhi　　　　安麟奎 An Linkui

基地调研 ‖ Site Research

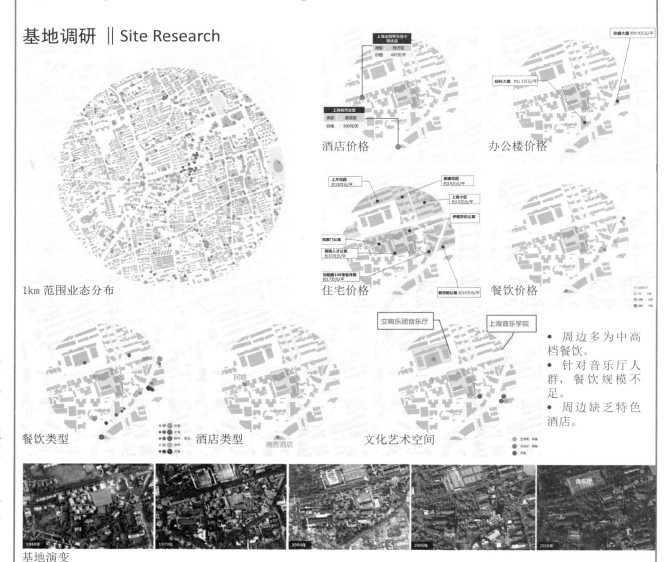

基地演变

- 街区：随着上海音乐学院的成立，琴行等音乐产业逐渐产生，街区音乐氛围逐步形成。
- 基地：随着新建筑的建造，黑石公寓原有花园被逐渐侵占。

既有建筑调研 ‖ Existing Building Survey

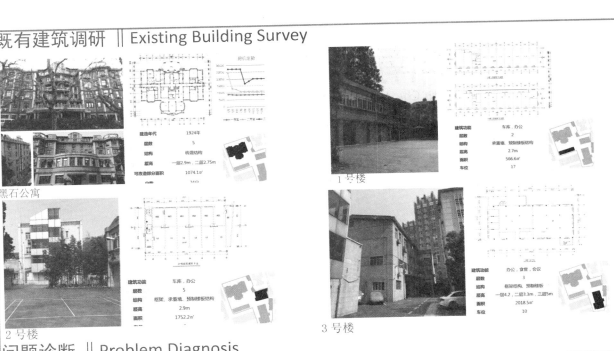

黑石公寓

1号楼

2号楼

3号楼

问题诊断 ‖ Problem Diagnosis

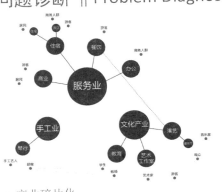

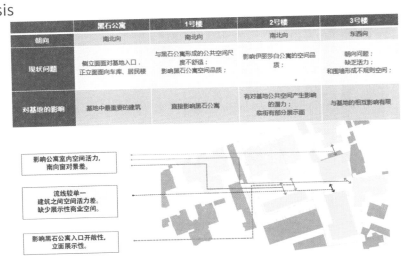

- 产业碎片化。
- 部分服务业缺乏特色。
- 文化产业缺乏持续运营的内在动力。
- 产业彼此分离导致人群之间缺乏联系。

概念提出 ‖ The Bringing Forward of Conception

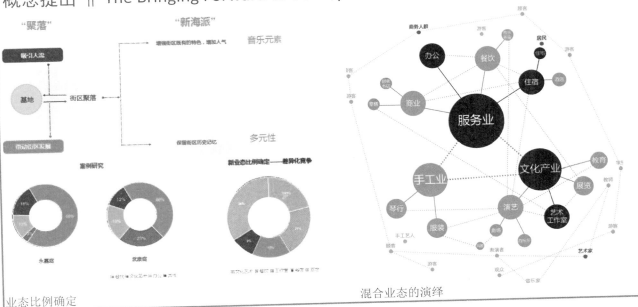

业态比例确定　　　　　　　　　　　　　　　混合业态的演绎

华南理工大学
South China University of Technology

CIID "室内设计 6+1" 2017（第五届）校企联合毕业设计
CIID"Interior Design 6+1"2017(Fifth Session)University-Enterprise Cooperative Graduation Design

一房一社区

小组成员：

林嘉辉 Lin Jiahui　　　　陈小帆 Chen Xiaofan　　　　卢惠兴 Lu Huixing
张裕麟 Zhang Yulin

概念构思 ‖ Concept of Design

 众

"众"这个概念可以引伸很多关键词。如众事、众聊、众知、众乐、众憩等，从而产生办公空间、社交空间、知识空间、玩乐空间、休闲空间。这些延伸出来的空间也正是我们对社区活动中心主要的空间属性。

从前，居民生活融洽，群体间不单是相互聚集，而且相互联系，是真正意义上的社区。

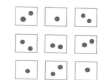
现在，随着时代发展。像上班族、留守儿童、独居老人产生，逐渐演变成仅仅居住在一起的独立群体。

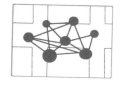
"众"是大家一起的意思，我们通过这个设计试图增加小区里各类群体的联系和交流，重新阐述社区的概念。

	众知	知识空间		
	众乐	玩乐空间		
	众事	办公空间		
	众联	社交空间		
	众憩	休闲空间		

國匠承啓卷——傳統民居保護性利用設計

建筑空间构思 ‖ Architectural Space Design Concepts

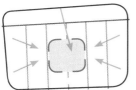
保留部分原有的建筑结构形成大的公共空间并串联其他空间。

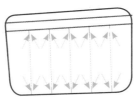
较完整地保留原有结构，促进空间交流。

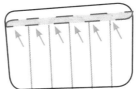
天井中绿植景观给人们带来了视觉上的享受。

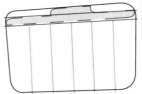
扩充天井增加使用者的公共活动空间。

原有的孤立乏味，相互隔离的单开间。

打破这个模式，提供更多的交流机会。

功能分析 ‖ Functional Analysis

由于打开了两个开间成为中庭体块，所以空间布局围绕中庭展开。原有空间仅靠一条直通的狭长楼梯进行楼层之间的连通，缺少空间之间的交流与联系。我们试图通过镂空楼层、设置观景阳台等方法，并在建筑外增加楼梯间和电梯间，连通上下空间，增强空间连续性，丰富了视觉感受和空间体验。

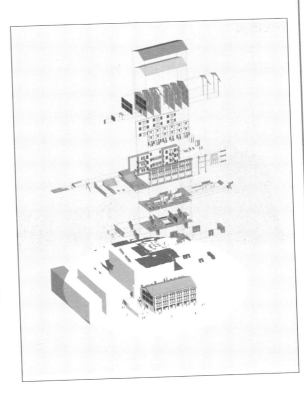

平面图 ‖ Floor Plan

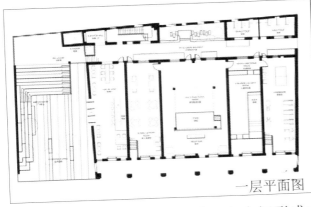
一层平面图

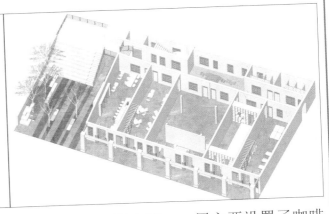

一层将部分旧墙体拆除，在中间形成大的活动空间汇聚人流。一层主要设置了咖啡厅（众憩空间）、老人活动室（众乐空间）、舞台（众乐空间）、多功能活动室（众乐空间）、儿童活动室（众乐空间）和健身室（众乐空间），以"众乐"和"众憩"空间为主。

哈尔滨工业大学
Harbin Institute of Technology

CIID"室内设计 6+1"2017（第五届）校企联合毕业设计
CIID"Interior Design 6+1"2017(Fifth Session)University-Enterprise Cooperative Graduation Design

里里·里院——哈尔滨老道外传统街区院落再利用与里院商业的探索

小组成员：

李　炎 Li Yan　　　　　　郭冰燕 Guo Binyan　　　　　　樊逸冰 Fan Yibin
刘名淞 Liu Mingsong

概念构思 ‖ Concept of Design

洋门脸　　vs　　中式内院
商业　　　vs　　居住
砖、石　　vs　　木

"里"的第一层次：内与外在建筑形式、空间功能和建筑材料的区别，形成对里院的准确认识。

但是，现在的商业模式已经发生改变

原有**前店后住**的商业模式，在现在已成为棚户区的**滋生地**。将**人流与商业**引入院落，形成新的**里院商业**。

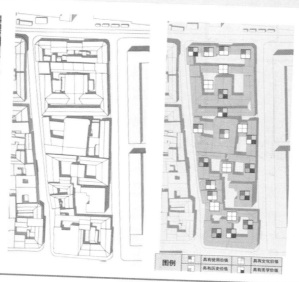

"里"的第二层次：根据建筑价值评估结果，对缺少保留价值的颓废建筑进行拆除，对具有使用、文化及历史价值的建筑进行保留。拆除建筑后，经过整理形成了一条贯穿里院的内街，由此形成里院商业。

场地现状

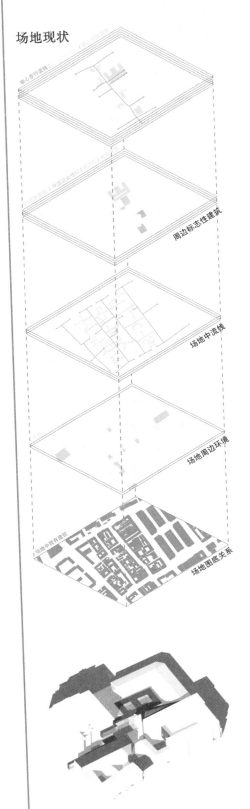

"里"的第三层次：选取街区内四个居民生活的活动中心，延续院落生活共享的特质，结合道外传统民居的院落装饰特点，应用于室内及景观设计当中。形成有别于哈尔滨道外中华巴洛克改造工程一期、二期改造的现代街区更新的新的模式，依照原街区居民生活中心，选取四个共享生活节点（新华书店、农业银行、省评剧院和松光电影院）作为里院共享商业节点，使整个街区内的院落间的沟通联系更为紧密。

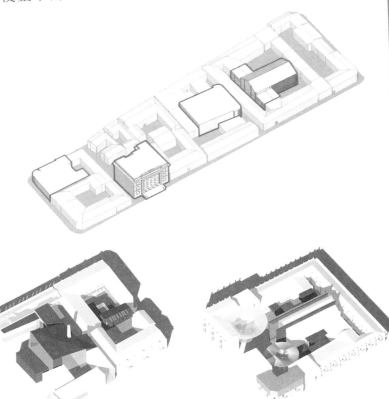

以新华书店为主的院落空间　　以省评剧院为主的院落空间　　以松光电影院为主的院落空间

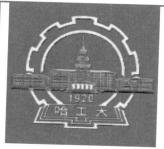

哈尔滨工业大学
Harbin Institute of Technology

CIID"室内设计 6+1" 2017（第五届）校企联合毕业设计
CIID "Interior Design 6+1" 2017 (Fifth Session) University-Enterprise Cooperative Graduation Design

市－井－里－院——哈尔滨市道外区中华巴洛克传统民居里院街区改造设计

小组成员：

胡一非 Hu Yifei　　　　单　昳 Shan Yi　　　　于　璐 Yu Lu

曹在笑 Cao Zaixiao

场地测绘 ‖ Site Survey

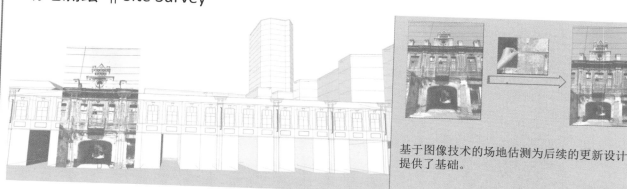

基于图像技术的场地估测为后续的更新设计提供了基础。

采访调研 ‖ Interviews

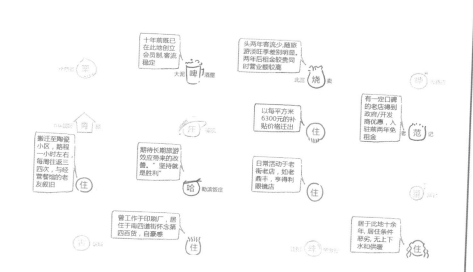

小组成员对16位场地上的居民和商铺的访谈，结论如图所示。

基地分析 ‖ Site Analysis

交通分析

景观分析

功能分析

保留价值分析

餐饮休闲区

文创体验区

购物展览区

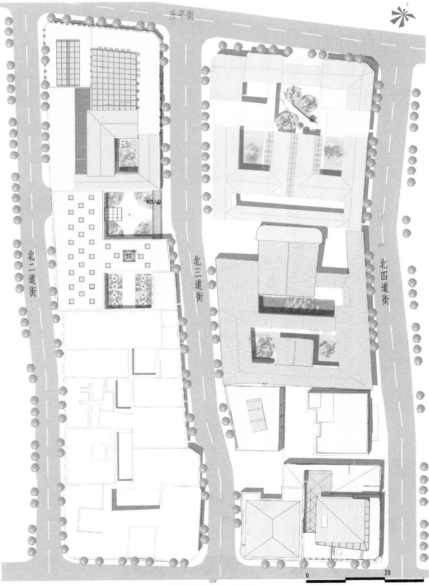
总平面图 1:1000

西安建筑科技大学
Xi'an University of Architecture and Technology

CIID"室内设计 6+1"2017（第五届）校企联合毕业设计
CIID "Interior Design 6+1" 2017(Fifth Session)University-Enterprise Cooperative Graduation Design

黄土印象——榆林市赵庄景观规划与建筑保护性更新及再利用设计

小组成员：

徐瑞显　Xu Ruixian	夏兆康　Xia Zhaokang	史文媛　Shi Weny
左怀腾　Zuo Huaiteng	杨　月　Yang Yue	田泽宇　Tian Zeyu

项目定位 ‖ Project Positioning

商业——窑洞书店特色商业街

旅游——凭眺全貌·层叠景观

文化——陕北窑洞文化的载体

打造一个立足于百年榆林张家大院的集商业、文化、旅游为一体的窑洞特色空间。以文化发展商业，以旅游带动商业，保护和开发相结合，为现代人营造一处回归文化的场所。

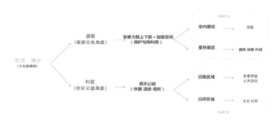

设计构想 ‖ Design Concept

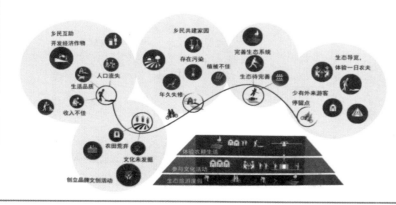

建筑上，窑洞坍塌、分布零散、公共空间的缺失；在环境上，水土流失、交通系统不完善、植被低覆盖率；在人文经济上，文化未挖掘、收入不佳、农田荒废。根据这些问题，结合当地实际情况，进行改造设计。

國匠承啓卷——傳統民居保護性利用設計

初步规划 ‖ Preliminary Planning

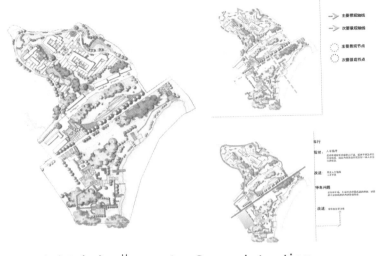

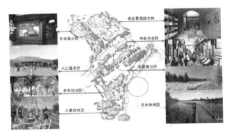

将整个基地进行基本功能划分，设计构想上突出陕北特色，为整个赵庄发展提供有力的支撑。

室内空间意向 ‖ Interior Space Intention

植入新型精品书店：在对建筑保护的前提下植入书店。提高建筑本身特质，情怀意境得到升华

特殊的营销方式：以经典藏书和人性化服务来打动读者，以文学界的活动，当地特色餐饮，读书为主题进行设计。

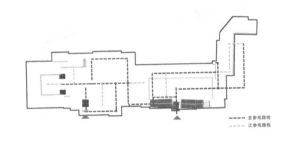

下院构想

上院构想

空外空间 创设场景

画廊的空间形态

上院构想

延续地域性文化营造，提取陕北传统符号，植入提高生活条件的现代元素，融入当地的生态元素，促成可持续的改造利用。让这书店成为有名景点，吸引游客，增加村民收入，改善人民的生活质量，为孩子们提供教育环境，让走出去的人们再回来，让村子又"活了"起来。

西安建筑科技大学
Xi'an University of Architecture and Technology

CIID"室内设计 6+1" 2017（第五届）校企联合毕业设计
CIID"Interior Design 6+1"2017(Fifth Session)University-Enterprise Cooperative Graduation Design

情系黄土　重返家园——陕州地坑院的传承与保护性设计

小组成员：

季　然　Ji Ran　　　　赵　涛　Zhao Tao　　　　崔为天　Cui Weitian

深入调研 ‖ In-depth Research

地坑院的优势与劣势分析

地坑院由于身居地下，并且门窗等材料的局限性，导致了采光效果极差，室内常年需要开灯来辅助采光。

地坑院的窑洞进深一般在 9～10 m 左右，进深偏大，在窑洞里待久了很容易有憋闷感，通风不畅。

地面的交通混乱，没有组织，没有明确的路网规划。村民的农机车辆基本处于乱停乱放状态。

建筑分析 ‖ Architectural Analysis

地下：近80~150年修建，保护程度一般，仍在使用。

地上：近20年修建，砖石结构，使用中。

地下：近100年修建，建筑格局完好，仍在使用。

地下：近100~200年修建，保护程度较差，已废弃。

地上路线分析 ‖ Land Line Analysis

地上的路线串联了全部的地上景观节点，又不破坏地坑院村落的原始样貌。每个地坑院都有道路可以到达，游览路线流畅。

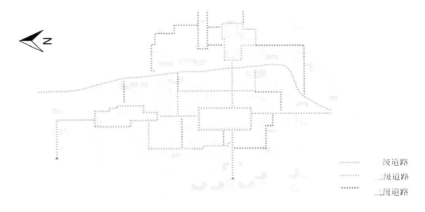

一级道路
二级道路
三级道路

地下路线分析 ‖ Subterranean Route Analysis

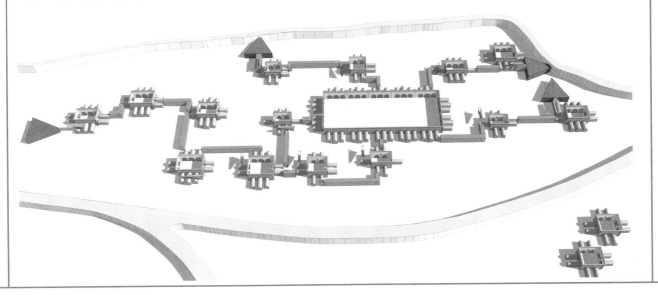

北京建筑大学
Beijing University of Civil Engineering and Architecture

CIID "室内设计 6+1" 2017（第五届）校企联合毕业设计
CIID "Interior Design 6+1" 2017(Fifth Session)University-Enterprise Cooperative Graduation Design

社区文化活动中心

小组成员：

汤博文 Tang Bowen　　　　　蔡明秀 Cai Mingxiu　　　　　刘佳蕊 Liu Jiarui

基地信息 ‖ Base Information

"家巧儿"社区活动中心位于北京市东城区北二环内青龙胡同西北角北邻雍和大厦等高层办公楼，南邻胡同居民房，西邻雍和宫柏林寺。涉及范围约24万㎡。活动中心场地面积约4800㎡，包含传统四合院一栋，戏楼一座，现代建筑一间，建筑面积共1240㎡。

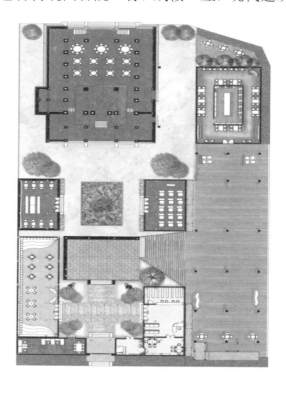

北京四合院传统营造技艺 || Beijing Courtyard Traditional Construction Techniques

二卷重檐悬山式

倒挂楣子

油饰

彩画

大木作彩画

木雕

戏楼概况 || Theater Overview

一层面积：568 ㎡。大厅面阔五间，进深七间，西面辟门板一间。

二层面积：328 ㎡。

舞台面积：45 ㎡。

坐北朝南，柱间面宽 5.68m，台宽 7.08m，台深 6.38m，柱间进深 5.68m。

85

南京艺术学院
Nanjing University of the Arts

CIID "室内设计 6+1" 2017（第五届）校企联合毕业设计
CIID"Interior Design 6+1"2017(Fifth Session)University-Enterprise Cooperative Graduation Design

甘熙宅第保护性利用设计

小组成员：

林秋霞 Lin Qiuxia　　　　　王依丽 Wang Yili　　　　　章伶钰 Zhang Lingyu

项目定位 ‖ Project Positioning

  本次毕业设计课题以甘熙宅第的保护性利用设计为主体，以拾遗——南京非物质文化遗产展为展览主题对其进行设计。总体设计原则宜游、宜承，让历史文化能衍生。基于甘熙宅第的保护规划，探讨设计范围内传统民居保护与利用的控制方式以及更新设计方式，创造有利营造技艺传承的展陈设计方式。

「融」
- 借原势形成新空间
- 因旧借新的光环境处理

"充分尊重
新旧相融"

「合」
- 展陈流线借势而行
- 展柜系统化零为整

"一体化"

处理六大空间关系

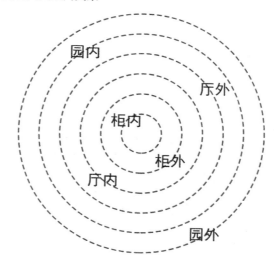

展览内容 || Exhibition Contents

南京拥有世界级非物质文化遗产四项，国家级非遗七项，另有省级非遗、市级非遗若干。我们从中挑选了三种，分别是南京云锦织造技艺（世界级）、秦淮灯会（国家级）、金陵刻经技艺（世界级）。结合展厅实际情况，进行古建筑的设计与改造。

展览规划 || Exhibition Planning

1　主入口
2　游客中心
3　小园
4　熙南里商业街区
5　津逮楼（藏书阁）
6　梨园雅韵（戏台）
A(1-4)　甘氏家族历史展示
B(1-5)　世界级非遗展示区
C(1-5)　金陵工巧（艺人工坊）
D(1-8)　国家级非遗展示区

南京艺术学院
Nanjing University of the Arts

CIID"室内设计 6+1"2017（第五届）校企联合毕业设计
CIID"Interior Design 6+1"2017(Fifth Session)University-Enterprise Cooperative Graduation Design

匠心椅阑干·甘熙宅第保护性利用设计

小组成员：

赵雨琦 Zhao Yuqi　　　　朱　彦 Zhu Yan　　　　黄雅君 Huang Yajun

项目定位　‖　Project Positioning

我们通过对甘熙故居的保护性改造利用，由中国坐具的工艺匠心作为切入口，向大家展示一个现代与古代、碰撞与传承、保护与发展的陈列空间。

将主题取名"匠心椅阑干"，将古典诗词中常用的"倚阑干"中的"倚"换做与主题相切合的同音字"椅"，意为传统技艺匠心的传承与期盼。

为了响应国家重点文物保护单位的要求，甘家大院在改造上有一定的局限。我们通过在周边商业区"熙南里"中增添一个新加建筑的方式，给甘家大院原有的散乱的出入口设立了统一的总入口，并在新馆前草坪上设立一个高大显眼的当代著名坐具装置作为标志。同时新馆内增添了票务、咨询、存包、导览、下午茶等功能性区域，使甘家大院的服务设施更加完整。

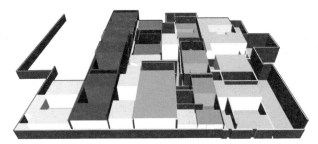

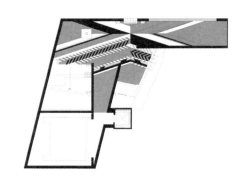

导向系统　‖　Guidance System

由于甘家大院本身有着院落繁多、行走路线杂乱的问题，我们设立了一整套的导向系统及标志符号。甘家大院是进递式的宅院，"进"的概念先入为主，我们将符号设立为形状与一进进排列的屋顶外形相似的箭头符号。既有甘家大院自身的特色，又有标识指向。

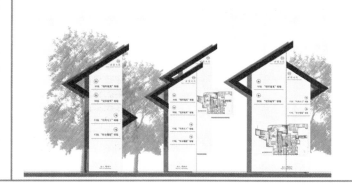

國匠承啓卷——傳統民居保護性利用設計

展陈大纲 || Display Content

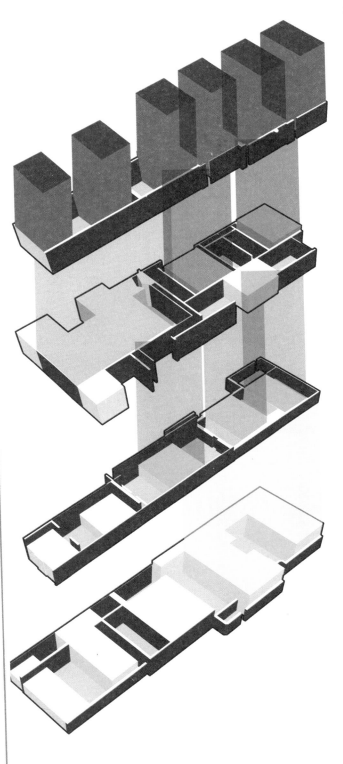

- 积厚流光

　　从精美明清家具回溯至古代时期，走近中国古代坐具发展史，从夏商周的榻到春秋战国至两汉的低矮家具，再到魏晋南北朝高形坐具的出现，再到隋唐五代直至明清的发展史简介（着重讲解明清坐具）以及介绍中国传统坐具与坐礼的关系以及变迁。

- 巧夺天工

　　现代家具手工制作人提供现场展示做法的空间。由古代家具匠人引入现代对工艺更对工艺人那份匠心的传承。

- 匠心独运

　　作为甘家大院内部现有最大的展厅，注重体验性，为参观者打造人人皆可参与的互动体验，最大程度上表现榫卯等坐具经典结构工艺的精细之美。

- 安居而坐

　　为参观者展示大量古代坐具实物，比如黑漆描金花卉纹交椅、核桃木灵芝纹嵌大理石扶手椅、红木嵌玉蝠纹宝座、榆木寿纹排椅、柏木高束腰圆凳等等。让参观者近距离欣赏中国传统坐具。

- 大工精诚

　　为参观者展示中国明清家具用材，主要有紫檀、黄花梨、鸡翅木、铁梨木、酸枝木、柞榛木、乌木、花梨木、樟木等。感受原材料的自然之美和椅子制作技艺的浑然天成。

浙江工业大学
Zhejiang University of Technology

CIID"室内设计 6+1" 2017（第五届）校企联合毕业设计
CIID"Interior Design 6+1"2017(Fifth Session)University-Enterprise Cooperative Graduation Design

半山村地块民居更新设计

小组成员：

高煊 Gao Xuan　　　　周睿 Zhou Rui

现状分析 ‖ Status Analysis

植物分析图

水文分析图

肌理分析图

原有规划图

半山村原有规划零散，村内交通也多为不便；村内有保留较好的古道、古庙、传统的竹编技艺等，村内竹海幽静，梨花壮美。

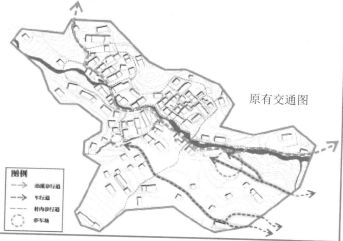

原有交通图

國匠承啓卷——傳統民居保護性利用設計

规划设计 ‖ Planning and Design

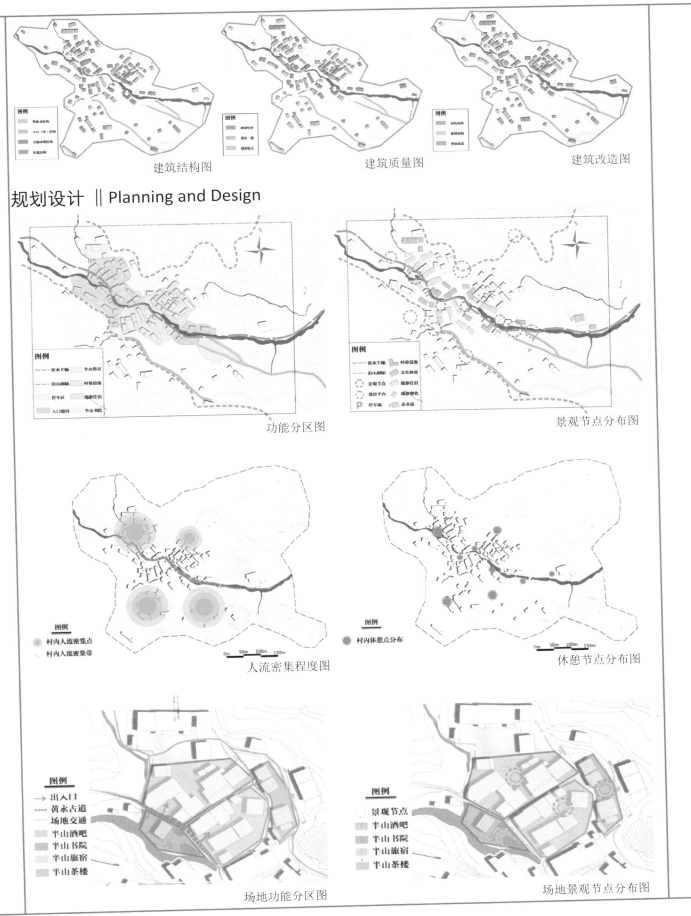

浙江工业大学
Zhejiang University of Technology

CIID"室内设计 6+1"2017（第五届）校企联合毕业设计
CIID"Interior Design 6+1"2017(Fifth Session)University-Enterprise Cooperative Graduation Design

浙江省台州市黄岩区半山村地块营造技艺传承与展陈设计

小组成员：

陈玫宏 Chen Meihong　　　胡安琪 Hu Anqi　　　洪朝艳 Hong Chaoyan
杨　洁 Yang Jie

区位分析 ‖ Location Analysis

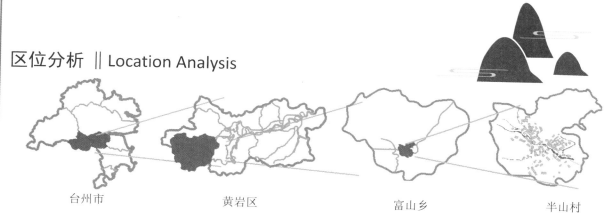

台州市　　　黄岩区　　　富山乡　　　半山村

　　半山村位于浙江省台州市黄岩区富山乡西南部，始建于北宋，盛于方和。与邻村相距五里，不上不下，故曰半山。群山环绕，竹林如海，梯田似锦，半山溪川村而过。

场地分析 ‖ In-depth Survey

现状分析

▲ 半山村业态分析　　　　　　　　　　　▲ 半山村业态分析

國匠承啓卷——傳統民居保護性利用設計

地方语言

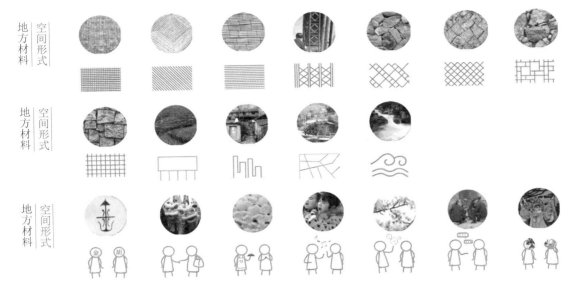

村民诉求

罗列半山村丰富的地方语言,将其重组,运用到设计地块的建筑空间和建筑内容上,并根据受众不同设计不同参观路径和参观方式。

概念生成 ‖ Concept Generation

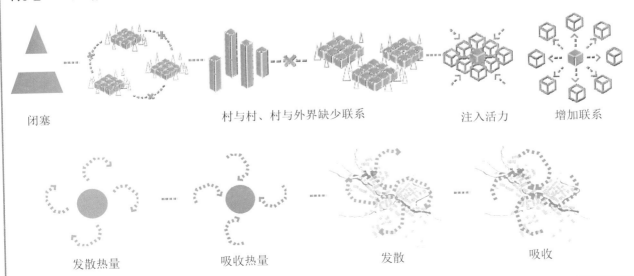

答辩展示
Defense Presentation

CIID『室内设计 6+1』2017（第五届）校企联合毕业设计
CIID "Interior Design 6+1" 2017(Fifth Session) University-Enterprise Cooperative Graduation Design

国匠承启卷——传统民居保护性利用设计
Craftsmanship Heritage
——Design for the Protective Utilization of Traditional Folk Houses

肆

國匠承啟卷——傳統民居保護性利用設計

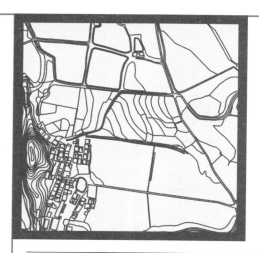

榆林市赵庄景观规划与建筑保护及更新设计再利用设计

CIID "室内设计 6+1" 2017（第五届）校企联合毕业设计
CIID "Interior Design 6+1" 2017 (Fifth Session) University-Enterprise Cooperative Graduation Design

高　　校：	西安建筑科技大学
College:	Xi`an University of Architecture and Technology
学　　生：	左怀腾　徐瑞显　夏兆康　杨月
Students:	Zuo Huaiteng　Xu Ruixian　Xia Zhaokang　Yang Yue
指导教师：	刘晓军　谷秋琳
Instructors:	Liu Xiaojun　Gu Qiulin
参赛成绩：	民居保护规划与景观设计二等奖
Achievement:	Second Prize for Folk House Protection Planning and Landscape Design

左怀腾
Zuo Huaiteng

徐瑞显
Xu Ruixian

夏兆康
Xia Zhaokang

杨　月
Yang Yue

学生感悟
　　四年的大学时光，马上就要结束了，而"6+1"联合毕设的答辩为我们画上了一个圆满的句点，经过近半年的努力，终于可以有一个交代。"6+1"联合毕设伴我们走完了大学的最后一段路，也是最精彩的一段路。在这段路上，我们跟6所学校的老师和同学共同的成长与进步。我们学习到了太多，不仅仅是知识。我们对于陕北民居从认识到改造设计，在老师的指导下一步步走来，让我们对设计本身有了更深的认识，感谢这次竞赛能给我们这么好的平台，也感谢老师与评委的谆谆教诲，我们将带着对设计的热爱之心，走向新的人生之路。

Students' Thoughts
　　As the four years' college life is coming to an end, our thesis defense for "6+1" joint graduation project has been completed successfully. We have made something of importance following diligent work over the past six months. "6+1" joint graduation project accompanied us to get through the last part of our college life, which was also the most spectacular moment. In this period, we developed and made progress with the teachers and students from six other universities. We've learned a lot other than knowledge. Thanks to the guidance of the mentor, we have had a deeper understanding about the dwellings in northern Shanxi from preliminary acquaintance to reconstruction design. We would like to express our appreciation to such a good platform providing us this chance to participate in the competition, and to the inculcation given by the teachers and judges. We'll step into a new life with our passion for design.

项目定位 ‖ Project Orientation

窑洞书店特色小吃街——商业
陕北窑洞文化的载体——文化
凭眺全貌·层叠景观——旅游

打造一条驻足于百年榆林张家大院的集商业、文化、旅游为一体的窑洞特色空间。以文化发展商业，以旅游带动商业，保护和开发相结合，为现代人营造一处回归文化的场所。

规划详解 ‖ Detailed Planning

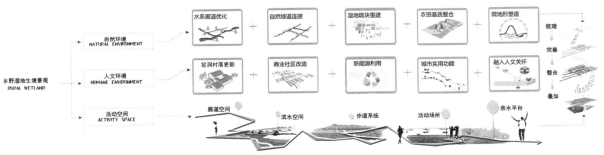

规划策略 ‖ Planning Strategy

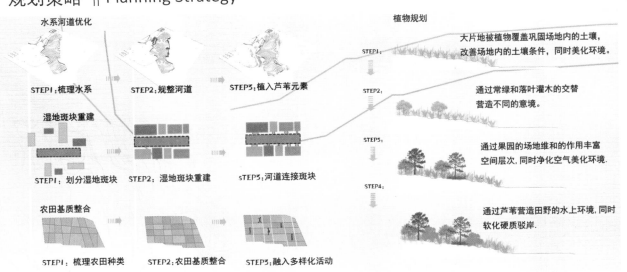

平面图 || Plan

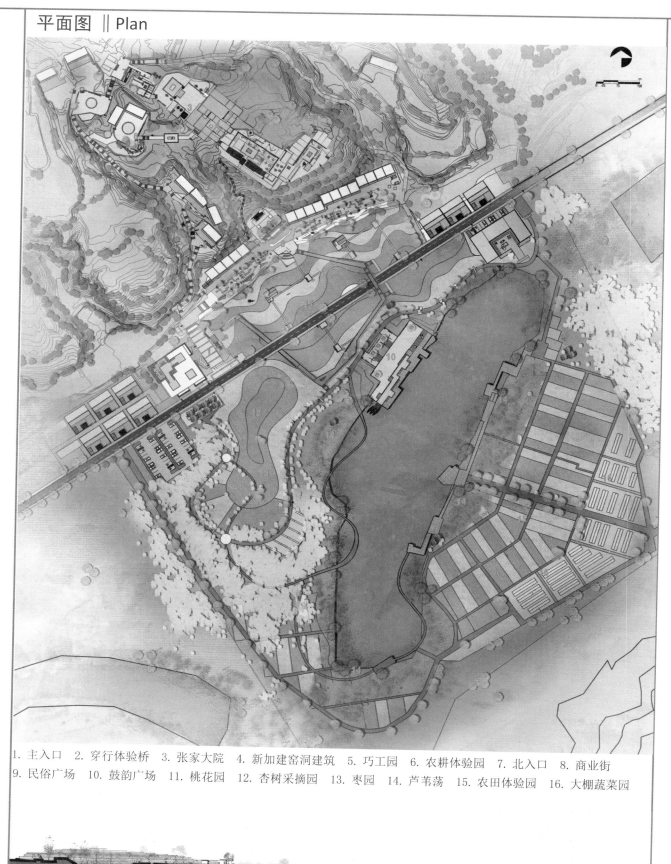

1. 主入口 2. 穿行体验桥 3. 张家大院 4. 新加建窑洞建筑 5. 巧工园 6. 农耕体验园 7. 北入口 8. 商业街
9. 民俗广场 10. 鼓韵广场 11. 桃花园 12. 杏树采摘园 13. 枣园 14. 芦苇荡 15. 农田体验园 16. 大棚蔬菜园

入口景观区 || Entrance Landscape Area

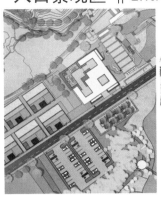

此区域为景区入口，设计以民俗建筑文化为主题，在设计中以窗花雕塑、陕北窑洞、民俗类的门墩、拴马桩、马槽、石碾子等进行造景，突出体现人文景观的魅力。

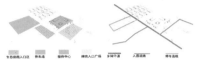

农耕园 || Agriculture Garden

农耕园设计目的为展示农事民俗农耕文化与农耕体验，

主要通过以"农"为主旨的景观石、农具犁、架子车、农具、石碾、马车、劳作场景、展示农事生产活动的文化景墙等，表现榆林的农耕文化。在基地的设计中最大的矛盾为交通干道使得基地南北方向一分为二，为解决这一问题，主要采用以种植形式和空中廊道来衔接场地。种植的形式以山体等高线演绎而成，空中廊道形式以小麦杆弯折形式演绎而成，农作物的种植以小麦、大豆、向日葵为主。

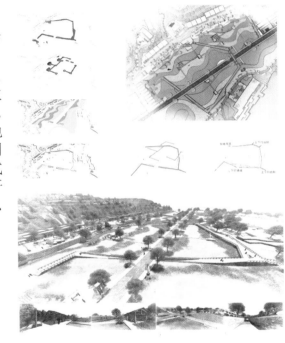

滨水生态区 || Waterfront Ecological Area

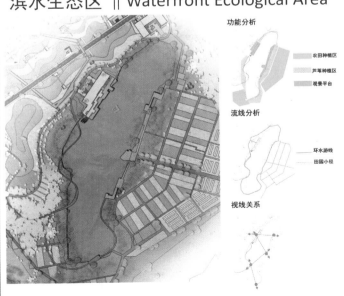

下院整体平面图 || General Plan of the House of Courtyard

整个下院院落主要以从新修复和利用为主，还原现破旧的建筑环境。在赋予新的功能空间的同时也不会破坏原有的空间感受和陕北风情。

功能及交通分析 || Function and Traffic Analysis

把废弃的窑洞修复，让过去曾在这里生活的人们回到这里，经营新的生意，增加经济收入，生活环境不变，生活质量大大增加。

院内窑洞空间新功能 || New Functions of Cave Dwelling Space in Courtyard

室内的空间根据设计的功能做装饰设计。材质家具，设计元素都是多用当地的传统家具元素，让空间有新的感受。

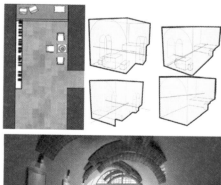

答辯展示・民居保護規畫與景觀設計

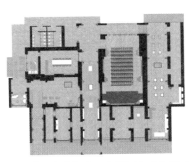

中院原本已经被岁月侵蚀的面目全非。我们把传统建筑和传统元素提取出来，利用当地建材设计了一个新建筑，把整个院落整体联系了起来。

到了上院,我们把原来院子里的破墙修复一半,把那种建筑的沧桑感留住让人观赏,把室内做成各种民俗体验和展示文化的空间。

导师点评

当今中国，城镇化与农村空心并存，生存环境的恶化和历史人文景观的消亡，成为社会日益关注的焦点。西安建筑科技大学的同学选择窑洞为设计研究对象，利用民俗艺术来装饰室内，使之古朴而时尚、实用而环保。在保持传统村落的自然与人文风貌的同时，又保护与传承了中国优秀民居文化传统。

朱飞

In today's China, urbanization and rural is coexistence, the deterioration of the living environment and the demise of the historical and cultural landscape have become the focus of social concern. Students choose cave for the design research object, the use of folk art to decorate the interior, so that simple and stylish, practical and environmentally friendly. To maintain the traditional village of natural and cultural style, while protecting and inheriting the cultural traditions of China's outstanding houses.

Zhu Fei

专家点评

设计紧扣"黄土印象"，充分挖掘黄土高原地形地貌及人文环境特征，结合赵庄现状山谷水库资源提出一系列合理的规划策略，营造出丰富且具有代表性的景观空间。对原有建筑进行修缮新建，并融入现代元素，营造出丰富的建筑空间形式。

黄土高原地区具有典型的地域特征，针对此区域的景观设计应该是典型的地域景观设计。沟壑纵横主要形成原因在于水与土的相互作用，如何留住土与水是设计的重点。应针对坡顶、坡面、庭院、小流域等不同部位进行深层次的研究，除了设计中提到的淤地坝还有人工梯田、鱼鳞坑、涝池、谷坊等多种收集雨水方式，可结合传统雨水收集方式营造出丰富且有意思的景观空间。同时，应针对陕北地域性植物进行更深层次的专题研究，找到合适的乡土物种，给出严谨可行的种植策略。在建筑处理上应挖掘对乡土材料、传统建筑布局形式、历史背景进行深入挖掘研究，营造出与黄土地真正相融合的乡土建筑。独特的环境给黄土高坡上的人们营造出了一种"慢生活"环境，独特的饮食、生活状态等恰恰区别于城市人，这也是黄土高原生活文化的魅力，应对此加以挖掘和展现。并且通过对黄土高原水系的治理，利用有限的水资源、加入生态环保理念来改善北方局部生态环境，达到吸引游人的目的。

总的来说，设计作品展现了大学四年中所学到的专业知识，也基本达到了高校教学培养的目的。伴随着专业知识的不断提升，希望同学对问题的研究分析越来越深刻，能透过现象挖掘本质，在设计行业越走越远。

寇建超

The design sticks to the theme of "Loess Plateau Impression" and fully taps the terrain and topography of the Loess Plateau and the characteristics of humanities environment. A range of reasonable planning strategies are proposed on the basis of the current reservoir resources in the valley of Zhaozhuang Town, and rich typical landscape space has been created. In addition, diversified layouts of architectural space are created through renovation and new construction of the original architectures with the integration of modern elements.

The Loess Plateau region has typical regional characteristic, therefore typical local landscape design shall be conducted for the region. The key point of the design lies in how to retain the soil and water as the interaction between soil and water is the main cause for the formation of the crisscross ravines and gullies. In-depth research shall be implemented on different sections, including the top and surface of slope, courtyard and small watershed. In addition to sediment storage dam for building farmland mentioned in the design introduction, various methods for rainwater collection, such as artificial terrace, fish scale pit, waterlogging pool and check dam are proposed, which may be combined with the traditional way of rainwater collection to create a rich and interesting landscape space. At the same time, more specially-purposed research shall be conducted on regional plants in northern Shaanxi so as to identify the appropriate local species and put forward a rigorous and feasible planting strategy. In addition, in-depth research shall be conducted on local materials, layouts and historical background of traditional architecture in respect of architecture design to establish a local architecture that is fully integrated with Loess Plateau. The unique environment has created an atmosphere of "slow life" for local residents lived in the Loess Plateau, where the special dietary habit and living conditions distinctively differ from that of urban residents. That also shows the charm of the living culture in the Loess Plateau, which should be excavated and demonstrated. For the purpose of attracting more tourists, the ecological environment of part of northern Loess Plateau are improved through the management of water system with utilization of limited water resources based on the ecological protection concept.

Kou Jianchao

社区文化活动中心

CIID"室内设计 6+1"2017（第五届）校企联合毕业设计
CIID"Interior Design 6+1"2017(Fifth Session)University-Enterprise Cooperative Graduation Design

高　　校：	北京建筑大学
College：	Beijing University of Civil Engineering and Architecture
学　　生：	汤博文　蔡明秀　刘佳蕊
Students：	Tang Bowen Cai Mingxiu Liu Jiarui
指导教师：	杨琳　朱宁克
Instructors：	Yang Lin Zhu Ningke
参赛成绩：	民居保护规划与景观设计组二等奖
Achievement：	Second Prize for Folk House Protection Planning and Landscape Design

汤博文　　　　　蔡明秀　　　　　刘佳蕊
Tang Bowen　　　Cai Mingxiu　　　Liu Jiarui

学生感悟

　　毕业设计是我们作为学生在学习阶段要经历的最后一个环节，是所学基础知识和专业知识的一种综合运用。是一种综合的再学习再提高的过程，这一过程对我们的学习能力和独立思考以及工作能力，也是一个培养。同时毕业设计的水平也反应了大学教育的综合水平，因此学校十分重视毕业设计的环节。在设计过程中，我们仔细聆听了老师的见解，增强了自己的领悟能力，了解了真正的设计流程。感谢6+1毕业设计给予我们的机会，让我们见识了很多学校同龄人的设计灵感和思路。受益颇深。

Students' Thoughts

　　Graduation project is the last link in our university study and a kind of comprehensive application of the basic knowledge and specialized knowledge we have learned. It is a comprehensive process of relearning and further improvement as well, and this process can cultivate our learning ability, independent thinking ability and work ability. Meanwhile, as the level of graduation project can reflect the overall level of university education, our university pays attention to graduation project. In the design process, we carefully listened to our teachers' views and tried to strengthen our comprehension ability. That is a real design process. We would like to thank "6+1" graduation project for giving us an opportunity to know the design inspiration and thinking of our peers from other universities. It can be said that we have benefited a lot from this activity.

屋顶农场有机餐厅

胡同农场 ‖ Hutong Farm

通过屋顶种植与可食地景相结合的屋顶农场有机餐厅解决绿地缺失（区域问题）、食品安全（社会问题）、居民种植需求（生活问题）等问题。

屋顶构造

- 种植土
- 排水层
- 过滤层
- 保护层
- 根阻防水层
- 找平层

750mm种植土
120mm蛭石排水层
过滤层：250~300g/m² 的聚酯无纺布
20mm混凝土保护层
3mmAPP聚酯卷材和3mm抗根卷材做防水层
20厚1:3水泥砂浆找平层
原建筑结构层

技术
太阳能　　　　　　　雨水收集循环系统　　　　　生态种植箱

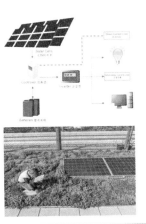

答辩展示·民居保護規劃與景觀設計

功能 ‖ Features

1. 室内功能

室内设有茶室、阅读室、学堂、酒廊、电脑室、医疗站、办公区。

2. 室外功能

健身器材区：为居民提供强身健体的器材，供居民自发开展文化体育活动。

广场区域：为胡同居民提供一个休闲娱乐的场所，可以晚间供居民跳广场舞。

屋顶露台：在不破坏传统坡屋顶的基础上，搭架起玻璃幕顶，既节省了空间，又为人们提供喝茶赏月的场所。

蔬菜种植区域：城市农场的理念融入其中使人们亲力亲为与大自然接触。

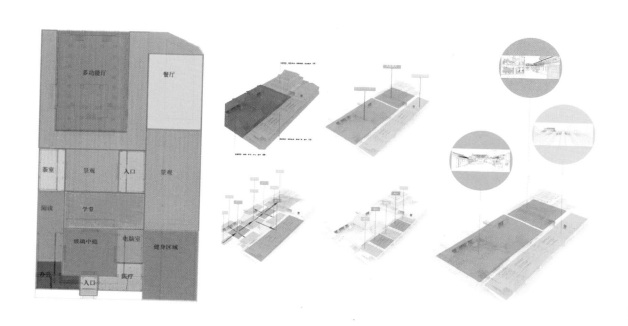

理念 ‖ Idea

- 健康的生活方式让我们不用逃离都市，可以更勇敢和创造性地面对生活。
- 有生命力的东西总有成长过程，所以不用急，慢慢来。
- 共生系统中，菜地为餐厅提供新鲜蔬菜，餐厅的果皮和咖啡渣又作为堆肥回馈菜地。
- 鲜嫩诱人的青菜苗，色泽艳丽的小西红柿，青翠欲滴的小黄瓜……
- 在种植的过程中，原本陌生的邻居在一起施肥浇水，分享种菜心得，不但种出了各色瓜果蔬菜，也种出了邻里间的感情，这是最大的收获。
- 透过耕作增加人与人、人与食物以及人与土地之间的互动。
- "城市农夫"，运动身体、舒缓压力，新鲜瓜果蔬菜，缩短农场到餐桌的距离，水泥森林回归自然。
- 即摘即食。

导师点评

四合院是华北地区传统民居的典型，也是北京历史风貌的代表，这个选题围绕城市化浪潮中旧民居转型利用设计的价值取向，其社会学层面上的意义超越了设计本身。

项目占地规模较大，场地条件复杂，建筑体量丰富多样，在空间组织、模型建构以及图纸表达上有较大难度。从设计过程看作者把注意力落在新旧建筑的形体协调上，希望在体量和形态上减弱新旧功能的矛盾。图纸内容丰富，版面整洁，总体设计表达清楚，体现了良好的基本素养。可惜在素材组织和设计内容上缺少对社区和人的关注，没能体现出重拾社区集体记忆的意图。另外受篇幅限制，设计内容覆盖新旧建筑多个空间，影响了设计重点的表达。

谢冠一

The quadrangle dwelling is typical traditional dwelling in North China and a model representing the historical style and feature of Beijing. The selected topic focuses on value orientation of the design adopted in the transformation of the old dwellings during the wave of urbanization. Its sociological significance is greater than that of the design itself.

The project covers a large area with complex site conditions and various sizes of buildings. Therefore, it is a tough task in terms of spatial organization, model construction and drawing presentation. It can be observed from the design process that the designer attaches importance to the harmony between old and new buildings, aiming to weaken the discrepancy between old and new functions through design of volume and shape of the buildings. Professional attainments are shown through the following aspects: The drawing is rich in content; the layout is clean and tidy; and the overall design presentation is clear. It is a pity that the design fails to present the intention with respect to regaining of collective memory in the community due to lack of concerns about the community and people when it comes to the material organization and design content. In addition, the design content covers various spaces of the old and new buildings because of the limited room for design, thus affecting the presentation of the key points of the design.

Xie Guanyi

专家点评

传统民居的围合式院落空间已经与现代生活难以匹配，设计者大胆引入一个雕塑般的构筑物，营造公共活动空间，增强了社区服务中心的开放性，为居民提供一个社交活动场所，打破了四合院的封闭形象。构架上部的种植唤醒阡陌沟渠的回忆，为社区提供一抹绿色，一股生机，更是孩童嬉戏、学习的最佳环境。不足之处是设计者未能通过构筑物对四合院进行充分的功能解构和流线分析，对于社区服务中心的多种功能及其布局缺少进一步的观察和思考。

张磊

The enclosing courtyard space of traditional dwellings is unable to keep pace with the modern life. The designer boldly introduces a sculptural structure to create a public space for activities, and enhance the openness of the community service center to provide residents with a place for social activities, thus changing the image of the old quadrangle dwelling closed to the outside world. Plants grown on the upper part of the architecture bring reminiscences of crisscross footpaths between fields, irrigation canals and ditches, provide the community with a touch of green and vitality, and more importantly, provide the best environment for the children to play and study. The deficiency of the design lies in that the designer fails to fully conduct functional deconstruction and streamline analysis toward the quadrangle dwelling through structures, and fails to do further observation and consideration of the various functions and layout of the community service center.

Zhang Lei

上海近代黑石公寓及其周边环境保护与更新设计

CIID"室内设计 6+1"2017（第五届）校企联合毕业设计
CIID"Interior Design 6+1"2017(Fifth Session)University-Enterprise Cooperative Graduation Design

高　　校：	同济大学
College:	Tongji University
学　　生：	张雨缇　周雨桐　谢丹妮
Students:	Zhang Yuti　Zhou Yutong　Xie Danni
指导教师：	左琰　黄全
Instructors:	Zuo Yan　Huang Quan
参赛成绩：	民居建筑保护与更新设计组一等奖
Achievement:	Frist Prize for Folk House Protection and Update Design

张雨缇
Zhang Yuti

周雨桐
Zhou Yutong

谢丹妮
Xie Danni

学生感悟

　　感谢在同济大学的五年，让我们收获了友情、理解、青春、乐观和团结。在这个2017年，从3月的开题，再到4月的汇报，最后6月的答辩，每时每刻我们的都心系着我们的毕业设计，更特别的就是我们跟6所学校的同学和老师共同走过了一段难忘的旅程，学习了更多更深的知识。给了我们一次难得的机会认识来自全国各地不同高校的小伙伴们和睿智的老师们。这次关于黑石公寓改造的命题，让我们对设计有了新的认识——设计不仅仅是用来解决建筑上的问题，更重要的还有解决周边策划、活力提升等等的问题。祝愿在毕业之后，每个人都能前程似锦！

Students' Thoughts

　　Thanks to five years' study in Tongji University, we started to know what are friendship, understanding, people, consideration, youth, optimism and solidarity. In the year of 2017, we experienced the establishment of our design topic in March, reported our work in April and completed our thesis defense in June, during which we never forget our graduation project. Especially we have got the unforgettable memories we experienced with teachers and students from other six universities and learned a lot from them. It is a rare opportunity for us to get acquaintance with the sagacious teachers and students from different colleges and universities around the country. We have a new understanding of design in relation to the project theme- Reconstruction of Blackstone Apartment- that the application of design cannot simply deal with architectural problems, but handle issues such as surrounding environment planning and architecture vitality promotion, etc. We wish that each of us have a splendid prospect after graduation!

國匠承啓卷——傳統民居保護性利用設計

场地分析 ‖ Site Analysis

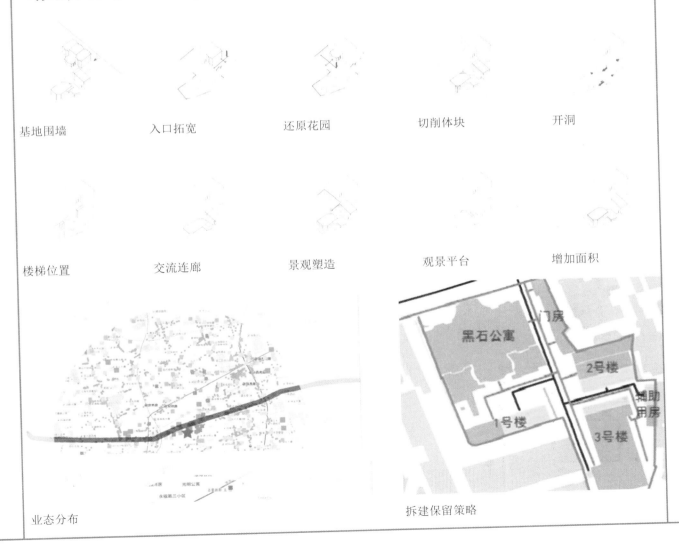

| 基地围墙 | 入口拓宽 | 还原花园 | 切削体块 | 开洞 |

| 楼梯位置 | 交流连廊 | 景观塑造 | 观景平台 | 增加面积 |

业态分布

拆建保留策略

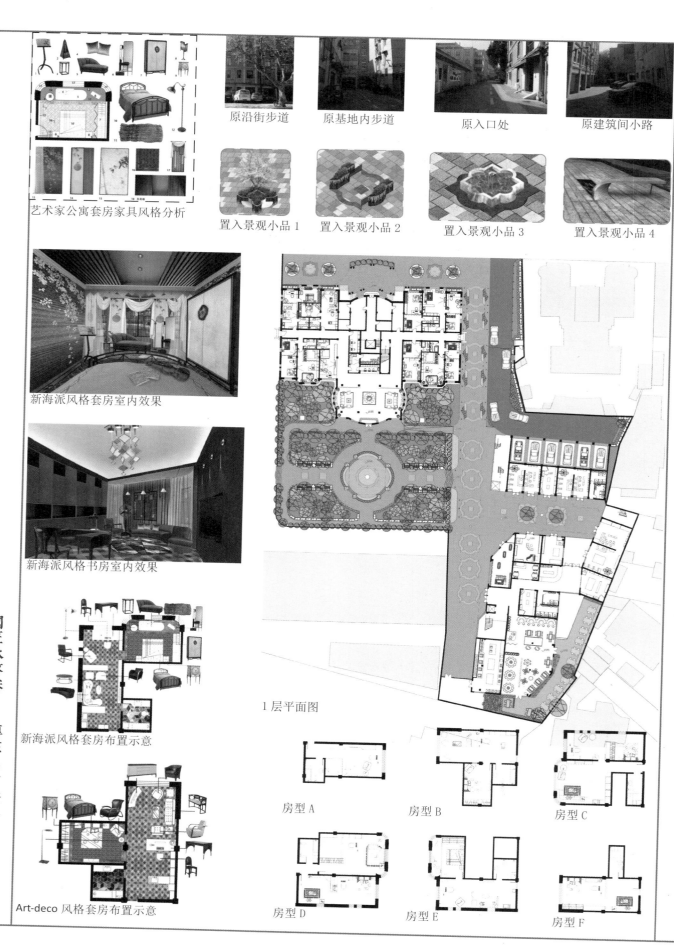

艺术家公寓套房家具风格分析

原沿街步道　　原基地内步道　　原入口处　　原建筑间小路

置入景观小品1　　置入景观小品2　　置入景观小品3　　置入景观小品4

新海派风格套房室内效果

新海派风格书房室内效果

新海派风格套房布置示意

Art-deco风格套房布置示意

1层平面图

房型A　　房型B　　房型C

房型D　　房型E　　房型F

國匠承啓卷——傳統民居保護性利用設計

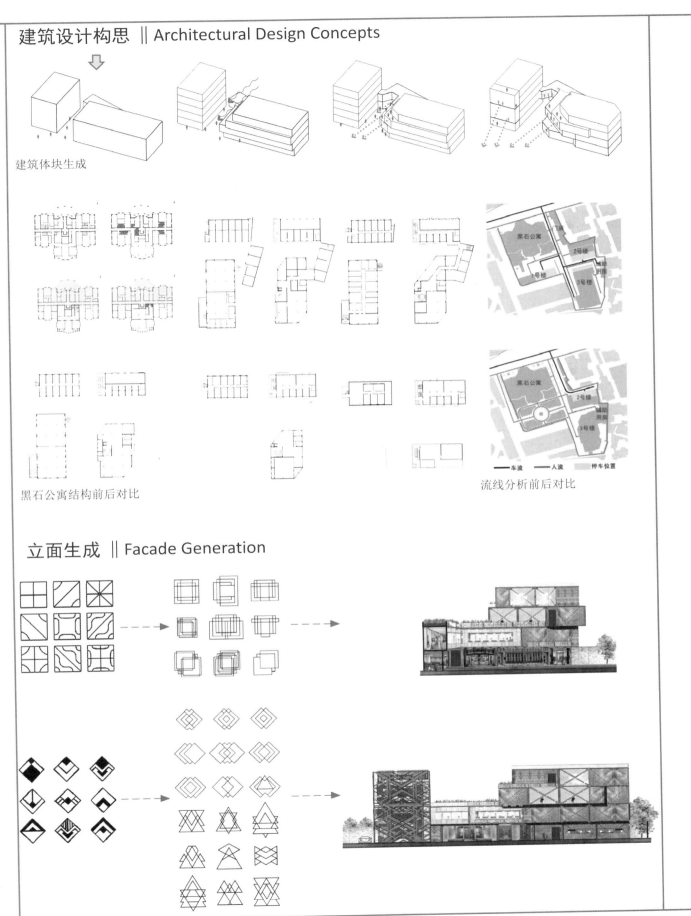

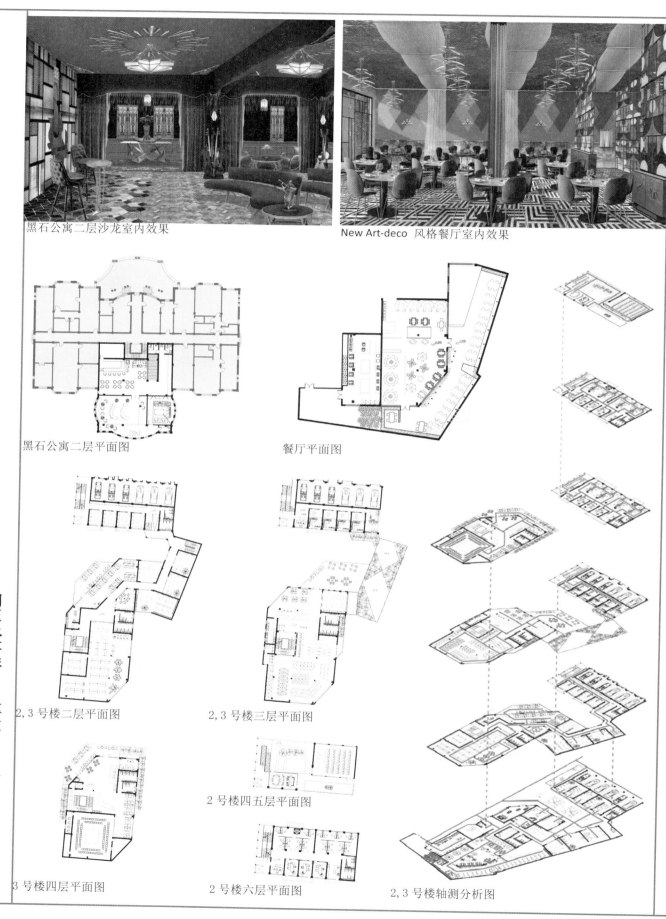

黑石公寓二层沙龙室内效果

New Art-deco 风格餐厅室内效果

黑石公寓二层平面图

餐厅平面图

2,3号楼二层平面图

2,3号楼三层平面图

2号楼四五层平面图

3号楼四层平面图

2号楼六层平面图

2,3号楼轴测分析图

國匠承啓卷——傳統民居保護性利用設計

导师点评

该组的设计方案以上海近代建筑黑石公寓改造为目标,对建筑周边环境进行了认真的调研。通过分析场地建筑肌理的演化和周边历史建筑的发展现状,确定了黑石公寓的保护价值和设计定位。并根据基地周边的业态,进行建筑的功能定位,较好的处理了建筑与街道、景观及相邻建筑的空间关系。设计思路清晰,技术得当,原真性的保护策略突出了黑石公寓的历史价值。在室内环境设计中,材质运用及装饰配置等都尊重了黑石公寓的历史文脉,形成典雅、时尚的新海派风格。配楼的设计自由、灵活、现代,与黑石公寓古典、庄重的气质相得益彰,主次分明。整个建筑更新方案设计理念统一,逻辑性强,运用手法恰当、实用,设计表达明快、精炼,表现出设计者良好的专业修养和综合解决实际问题的能力。

周立军

The team takes reconstruction of Blackstone Apartment, a modern building in Shanghai, as the objective of their design proposal, conducts a conscientious investigation on surrounding environment of the architecture, determines the protection value and design orientation of the Blackstone Apartment through analysis on the evolution of architecture texture on the site and the development of the surrounding historical architecture. In addition, the team arranges functional orientation of the architecture based on the type of operation around the foundation, manages the relationship between the architecture, street, landscape and adjacent buildings in a good manner. The clear design thought, proper techniques and the originality and authenticity of the protection strategy, all highlight the historical value of the Blackstone Apartment. In terms of the interior environment design, the historical context is respected in terms of the material application and decoration arrangement, thus creating a new Shanghai-style building with elegance and fashion. The free, flexible and modern design in annex building is able to complement with and highlight the importance of the classical, solemn Blackstone apartment. The updated design project for the overall architecture is characterized by uniform design philosophy, strong logic, proper and practical application approach, and lucid and refined design expression, indicating the professional knowledge and comprehensive capability of the designer for resolving problems.

Zhou Lijun

专家点评

黑石公寓项目,属"旧建改造及活化"性质。其设计成果的特点在于:

(1)为"活化"而梳理与整合的业态明晰完整——因应毗邻的上海音乐厅,在本项目内架构起层次分明相互呼应的音乐消费链。

(2)为"改造"而归纳出的问题链,以及对应"链"上诸问题而采取的改造路径准确并有效——例如"拆"1号楼"连"2号楼"裁"3号楼;以及"分"——人车分流、"融"——原有造型母题的融合塑造等。

(3)为"提升"而选取的设计逻辑清晰且流畅——例如针对历史片段与当下衍变而确定的联接手段、针对凹陷区域的"里、外"代偿意识、针对老旧户型的柔性修辞手法。

(4)明晰的经济指标控制,扼要的可移动要素配置。

综上所述,黑石公寓项目应属本次"6+1"活动中的最优秀设计之一。

赵健

Blackstone Apartment project falls into "reconstruction and activation of the old architecture".

The design achievement of the project is characterized in the following:

(1)The distinct and full types of operation are arranged and integrated for the purpose of "activation"—a well-arranged and music consumption chain with concerted action is established under the project in response to the adjoining Shanghai Concert Hall.

(2)Problem chains are identified and summarized for construction, and various problems concerning the corresponding "chains" are handled through accurate and effective approaches. For example, "demolition" of Building No.1, "connection" to Building No.2, "cutting" of Building No. 3; and "separation" -- separation of pedestrians and vehicles, "integration"—integration and shaping of the original modeling and motif etc.

(3)The design logic selected for "promotion" is clear and smooth, for example, the connection method determined aiming at historical fragments and the changing world; "internal" and "external" compensation awareness for sunken area and flexible rhetoric aiming at old buildings.

(4)Purposeful control of economic indicators; compendious configuration of mobile elements.

To sum up, Blackstone Apartment project shall be regarded as one of the best design work in the 6+1 design event.

Zhao Jian

情系黄土 重返家园——陕州地坑院的传承与保护性设计

CIID"室内设计 6+1" 2017（第五届）校企联合毕业设计
CIID"Interior Design 6+1"2017(Fifth Session)University-Enterprise Cooperative Graduation Design

高　　校：	西安建筑科技大学
College：	Xi`an University of Architecture and Technology
学　　生：	季　然　赵　涛　崔为天
Students：	Ji Ran　Zhao Tao　Cui Weitian
指导教师：	刘晓军　谷秋琳
Instructors：	Liu Xiaojun　Gu QiuLin
参赛成绩：	民居建筑保护与更新设计组一等奖
Achievement：	Frist Prize for Folk House Protection and Update Design

季　然　　　　赵　涛　　　　崔为天
Ji Ran　　　　Zhao Tao　　　Cui Weitian

学生感悟

　　如果说毕业季充满了即将步入社会的紧张和师生分别的感伤，那么这次的"6+1"联合毕设之旅便让我们有一种新的情绪和感受：从3月的开题相聚，再到4月的汇报熟知，最后6月的答辩。我们跟6所学校的同学和老师共同走过了一段难忘的旅程。我们相信必定是怀揣着共同的目标和理想才让彼此有机会相聚。这次传统民居保护与利用的命题，也让我们对传统民居设计有了新的认识——传统民居所具有的精神与灵魂值得被传承下去。此次毕业设计有遗憾但更多的是成长，是结束同样也是开始。

Students' Thoughts

If the season of graduation is filled with nervousness for us to step into the society and a sentimental mood for farewell to teachers and schoolmates, then the "6+1" joint graduation project brings us a new sentiment and feeling. We had a memorable experience with the teachers and students from six universities since we met in March, when our design topic was established, and then we get acquaintance with each other when reporting our work in April and conducting thesis defense in June. We firmly believe that it's the common aspiration and ambition that have brought us together. Based on the project theme for the protection and utilization of the traditional dwellings, we have also had a new understanding of the traditional dwelling design, i.e. it is worth inheriting the spirit and soul of the traditional dwellings. Although our graduation project is not perfect, we learned a lot. The completion of the project also signifies that we still have a long way to go.

项目定位 ‖ Project Orientation

改善住居环境

带动旅游业

经济收入提升

年轻劳动力回归

可持续发展

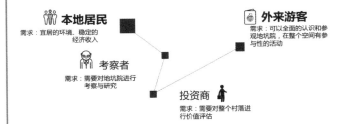

- 改善居住环境，解决通风、采光、卫生间等问题。延续下沉式窑洞的居住形式。
- 重新规划设计地面的窑顶空间，将其充分利用。完善村落道路系统和其他配套设施。
- 村民收入低，以农业种植业为主，希望通过设计提高村民收入。

设计详解 ‖ Detailed Planning

由于地坑院本身空间的局限性，会导致部分展厅的展览流线中断，以及在居住空间中的不便，例如储藏窑和厨窑是分开的，所以在设计中将部分的窑洞进行打通串联，增加了空间的丰富性，在展厅空间保证了浏览路线的通畅，在居住空间增加了空间的利用性。

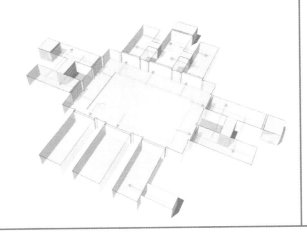

平面图 || Floor Plan

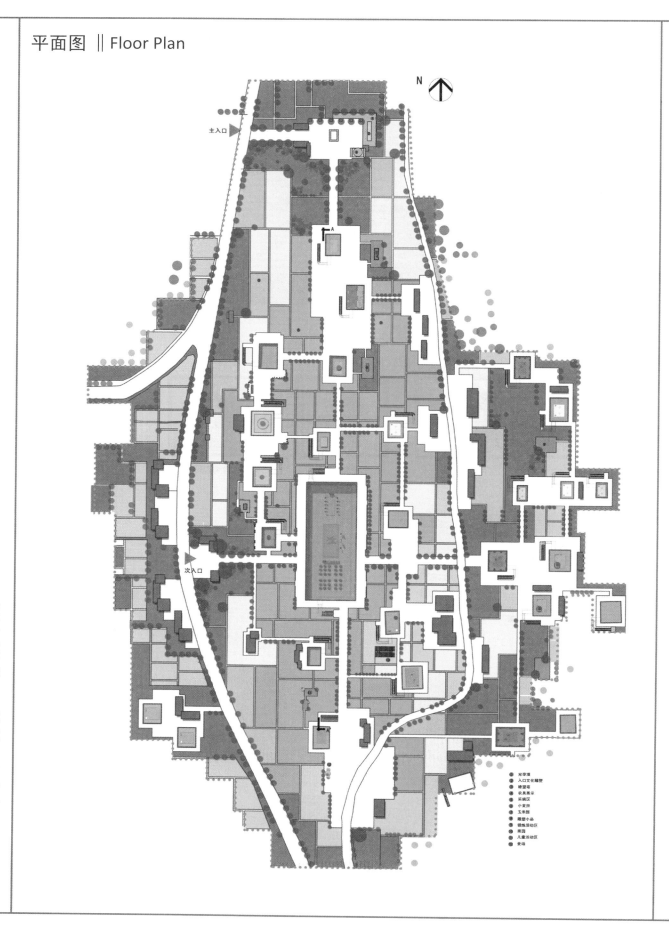

改造构思 ‖ Concept of Alteration

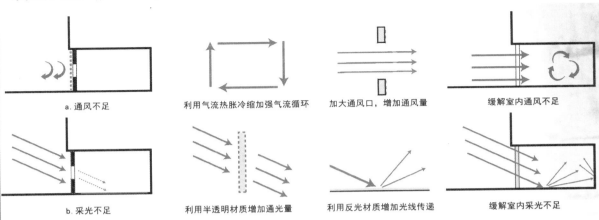

传统的天井式窑洞居住室内的通风和采光十分的不足，使得人在里面的生活舒适度直线下降，由于室内的通风与采光不足，使得窑洞内部气流不流通，极易发霉。因此希望通过加入现代材料与运用物理常识，在不破坏原有建筑形式的前提下，改善解决问题。为更好为窑洞使用者提供便利，在其中也尽可能解决如洗浴、上厕所等传统上无法提供的方面。

入口空间 ‖ Entrance Space

可以登上村口瞭望塔，展望整个村庄的景观，远眺就可以看到远处层层叠叠的塬上风光。

入口荣誉墙

渗水处理 || Seepage Treatment

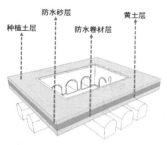

窑顶的"双层夹砂"营造技艺

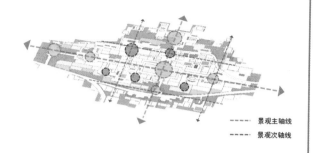

地坑院的窑顶之前由于营造工艺的不足，窑顶易产生渗水等现象，威胁窑洞的安全，技术改进后采用"双层技术"，在土中夹砂，可以有效蓄水和防止水的下渗。窑洞空间就被充分利用起来。

功能分区 || Functional Division

总平面经过规划以后分为12个功能分区，分为主次入口广场、历史展示区、文化展示区等。以地坑大院为中心集散广场。后边为民宿居住区，供本地居民居住，并且提供民宿。

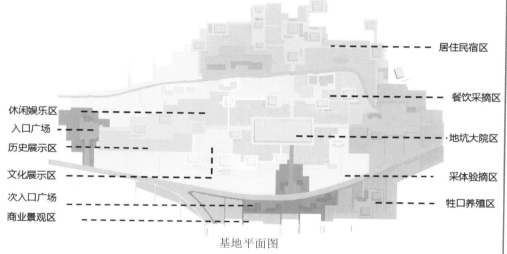

基地平面图

展示区 || Display Area

历史展示区展示了生土建筑的建造过程及历史文化。使参观者直观感受生土建筑的厚重文化。

文化展示区的展厅全部打通，可以完整的参观完整个地坑窑洞，保证路线的通畅。

地坑大院总平图

地坑大院室内效果图

居住民宿区 ║ Living Guest House

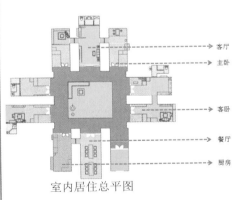

室内居住总平图

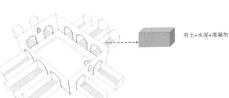

地面工艺改造

在地面工艺的改造上，使用夯土加入水泥以及混凝剂。首先加入黄土，保留了窑洞中的黄土记忆。水泥和混凝剂增加地面的坚固性，防潮性也大大增加。

主卧效果图

居住区效果图

居住民宿区的设计的主要目的是改进当地居民的居住环境，吸引游客前来体验原汁原味的地坑院生活，将卫生间等空间引入地下。

导师点评

本设计以黄土高原的传统居住形态为背景，目的在于通过现代的设计理念，将传统居住空间重新利用起来。"情系黄土，重返家园"抒发了学生对家乡的情怀，旨在唤起在外游子们的那份"乡愁"。整体规划布局在主轴线的统领下，通过完整的路径设计将各个功能片区整合起来，再加上细致的室内设计，将传统和现代共同演绎在一个传统的室内空间中，既尊重了历史，并对原有的空间格局进行了重构。设计深入，具有很强的可行性。

左琰

Based on the background of traditional dwellings in the Loess Plateau, the design is designated to reuse the space of the traditional dwellings through modern design concept. "Affection of Loess Plateau, Returning Home" gives expression to the affection of the students towards their hometown, aiming at arousing "nostalgia" of the people travelling or residing in a place far away from their hometown. The overall planning layout make a good arrangement for both traditional and modern elements within one traditional interior space by adding exquisite interior design with integration of various functional areas through a complete path design under the guidance of the main axis, indicating respect for history and reconstruction of the original spatial pattern. The work is designed deeply with a good feasibility.

Zuo Yan

专家点评

设计围绕"情系黄土，重返家园"的设计理念对地坑窑这一古老生土建筑形式进行研究，梳理建筑脉络，并注入新的业态形式，重建景观轴线，将没落的建筑群进行复兴和重塑。尤其是对地坑窑空间的采光改造及结合现代工艺对地坑窑环境的改良尤为突出。整个设计思路连贯，表达尤为精彩。

地坑窑作为一种古老而神奇的民居样式，蕴藏着丰富的文化、历史和科学，被称为中国北方的"地下四合院"具有极高的研究价值。"山有去脉，水有流向，土有层纹"营造地院要与山脉、水势、地气相融合，要以五行八卦和主人命相来定方位。地坑窑是风水学、建筑学、人文学的绝妙结合，是传统哲学思想"天人合一"的现实体现。相院、方院、下院、打窑、绿化……都有严格的方位、尺寸、数量要求和禁忌，窑洞的功能设置，饮水、排水都有科学合理的布局法则。地坑窑已然成为一种文化符号，既然是"重返"，就要对地坑窑形成的原因、营造技艺进行深层次的挖掘才能指导我们科学地保护和修复，此次设计中对此研究不够深入，很多东西还浮于表面。同时，在景观节点的营造上，应结合乡土材料展现乡土景观。设计中所体现的景观环境地域性不够明确。在文化展示方面，除了民俗，地坑窑就是活化石，可增加其建筑学、人文学、风水学的展示来丰富空间的历史文化内涵。在室内设计中，此次设计融入多种现代元素，十分丰富，但主题不够贴切。总体来讲，此次设计思路清晰明了，形成了相对完整的研究体系，充分展现了大学四年中所学到的专业知识，达到了教学培养的目的。希望同学在今后的能够更加增加研究的深度和广度，呈现出更精彩的作品。

寇建超

Based on the design concept of "Affection of Loess Plateau, Returning Home", research is conducted on the ancient earth construction form- underground cave dwelling in this design, where the downfallen architecture complex are revived and remodeled through landscape axis reconstruction with combing of architectural skeleton by injecting into new types of operation. It shall be mentioned that the particularly outstanding points for the improvement of underground cave dwelling environment rest with the lighting reconstruction for space of underground cave dwelling with incorporation of modern crafts. The overall design thought is coherent and its expression is especially remarkable.

As one ancient and magical style of architecture for residence, the underground cave dwelling is rich in culture, history and science. It is called "Underground Quadrangle Dwelling" with high research value in northern China. The mountain has its range, the water has a flow direction, and the soil has the laminated striation. The mountain range, flow of water and the soil properties shall be considered and incorporated into the construction of the underground cave dwelling located through the Five Elements and Eight Diagrams and the fate of the house-owner. The underground cave dwelling, takes advantage of theory of Fengshui, architecture and humanities concerning its construction, is the real entity expressing the traditional Chinese philosophy of "man is an integral part of nature". During the construction process of underground cave dwelling, including location identification and construction planning, positioning through timber piles, patio formation, cave dwelling digging and landscape engineering, strict requirement and restriction shall be conducted for positioning, dimension, quantity. The function setting, water supply and drainage are furnished based on scientific and rational law. The underground cave dwelling has become a cultural symbol. Since "returning home" is incorporated into its design concept, an in-depth research shall be conducted on the reason for creation of underground cave dwelling and the construction technique thereof in order for us to protect and renovate such architecture in a scientific manner. However, in view of the design work, relevant research was not conducted so deeply that the essence in many aspects was not perceived through the appearance. In addition, local materials shall be used for the construction of landscape nodes of the underground cave dwelling with a view to expressing local landscape. Locality of the landscape environment is not expressed clearly by the design. In addition to folk custom, the underground cave dwelling is considered as the living fossil in terms of culture presentation. By more utilization of the theory of Fengshui, architecture and humanities, more features of the underground cave dwelling will be shown to enrich the historical and cultural connotations of its space. While a wide range of modern elements are incorporated into this design in respect of interior design, the design work is not closely linked to the design theme.

Kou Jianchao

老城新生

CIID"室内设计 6+1"2017（第五届）校企联合毕业设计
CIID"Interior Design 6+1"2017(Fifth Session)University-Enterprise Cooperative Graduation Design

高　　校：	华南理工大学
College：	South China University Of Technology
学　　生：	郭嘉敏　谈　志
Students：	Guo Jiamin　Tan Zhi
指导教师：	石　拓　谢冠一　薛　颖　骆　雯
Instructors：	Shi Tuo　Xie Guanyi　Xue Ying　Luo Wen
参赛成绩：	民居建筑保护与室内设计组二等奖
Achievement：	Second Prize for Folk House Protection and Update Design

郭嘉敏　　　　　　谈志
Guo Jiamin　　　　Tan Zhi

國匠承啓卷——傳統民居保護性利用設計

学生感悟

如果说毕业季充满了即将步入社会的紧张和师生分别的感伤，那么这次的"6+1"联合毕设之旅便让我们有一种新的情绪和感受：从3月的开题相聚，再到4月的汇报熟知，最后6月的答辩。我们跟6所学校的同学和老师共同走过了一段难忘的旅程。我们相信必定是怀揣着共同的目标和理想才让彼此有机会相聚。这次关于居住环境改造的命题，也让我们对设计有了新的认识——社会问题和社会需求才是设计者的设计动机和动力。此次毕业设计有遗憾但更多的是成长，是结束同样也是开始。路漫漫其修远兮，吾辈将上下而求索！

Students' Thoughts

If we say the senior year is full of tension for entering the society and sadness for teacher-student separation, then this "6+1" University-enterprise Cooperation Graduation Project Activity has made us generate a kind of new feeling. We gathered in March for proposal, got familiar with each other in April for report and make oral defense in June. We had an unforgettable time with the teachers and students from six universities. We believe it was the common goal and ideal that allowed us to have the opportunity to gather together. The proposition about residential environment reconstruction has also made us have new understanding – social issues and social needs are the motive and impetus of designers. This graduation project has some problems but it has grown to a greater extent. It is the end as well as the beginning. We will keep exploring and advancing on the long road ahead of us.

答辩展示·民居建筑保护与更新设计

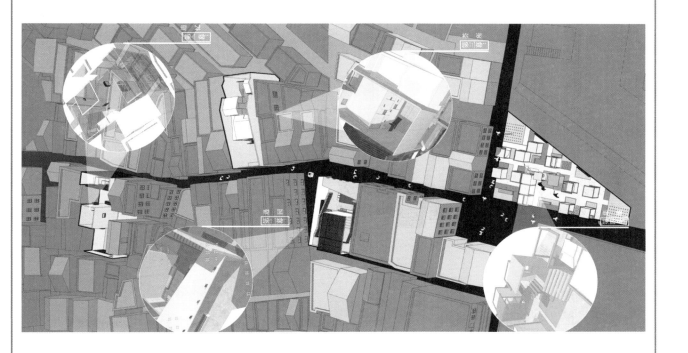

广州荔湾区泮塘五约村，一个饱含岭南传统文化底蕴的古村，正随着时间流逝渐渐被遗忘，沦为脏乱差的城中村。

我们尝试通过对这一地区的居住模式、生活模式进行一次更新改造，引入新的适应人群，为该地区带来新的经济效益和社区活力。

我们以五约直街作为具体的设计对象，以"极窄的生存，交通空间"为线索，探索创造"共享家"的老城社区生活模式。我们认为，"共享家"的老城社区生活模式应该由生活-办公空间、公共景观空间、交易景观空间、精神空间四个部分组成，并选择五约直街的四个现有建筑，经过更新改造形成四个这样的空间，使之自然地"生长"在老城社区中。

建筑空间构思 ‖ Architectural Space Design Concepts

基地本身特有的窄小狭长的主街形态

针对如何整合现有密集排布的竹筒屋，将单一的竹筒屋建筑相互之间产生联系性的问题。我们希望通过植入新的空间体块，利用连接与互动的形式，将房屋之间串联起来，达到共享的目的。以竹筒屋群为基点，在一定辐射范围内，承载满足年轻人群和内部居民的行为活动，创造公共到私密的新的秩序。

设计策略

直街窄小狭长，两旁建筑密集分布

室外空间延伸到室内，形成一个"凹"空间

多个"凹"空间和半开放式庭院串联，围合社区气场，成为区域的视觉焦点

村中现存的大量没有被利用的建筑

注入不同功能的建筑＋景观

把建筑＋景观扩散到整片区域

第一幕：追忆空间 ‖ The First Act: Memory Space

从空间上，它选址位于泮塘五约村直街上重点保护历史建筑旁，而从时间上，广州城区的转型和加速发展又将它推到这一新旧时间的交汇点上。藉此展开了如何将这一块已破损的传统民居的改造思路，而并不是将其视为一个孤立的事件。改造后，其外在形成泮塘五约村景观中的一个艺术装置，而它的内在将为周边社区提供一个新的公共活动空间——追忆展廊。

在日常的使用功能上，它还可能成为具有某种精神性的场所——精神堡垒。改造是在对待历史与现实的审慎态度下展开的，工业吞并传统，传统正在被新的传统替代，故将现代格局突破传统，并探索如何将新的现实植入历史样本中。一方面，尽可能不去触碰已破损的竹筒屋结构，只进行必要的结构加固和局部处理天井本身原有的窗洞口，进入后，几乎一眼望不到底的长廊，深处开窗进入的光，似乎喻示着传统光下的一把座椅，是传统，也是回忆。

空间演变

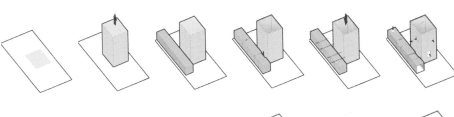

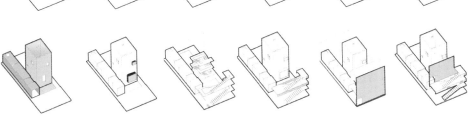

效果图

一部分精巧的装置被植入到天井内部，较小的位于天井顶部收集天光，较大的在天井底部形成了一个拉长的纵深空间，并呼应着多个采光窗——这些将环境光线经过间接反射导入天井内部的采光窗从天井每一个可能的开口"生长"出来。

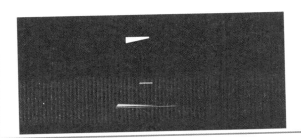 

答辩展示·民居建筑保护与更新设计

第二幕：遇园空间 || The Second Act: The Garden Space

此处为泮塘五约村的露天剧场，剧场顶部并非全封闭设计，透漏有序，使剧场有充分的光感体验，另起到遮阴效果。在这个半围合空间内，可以举行小型文娱活动以及唱粤剧。主入口临近直街，连接入口的通道既可以到达露天庭院也可以直接到达露天剧场，或者直接从临街小巷到达剧场内部。露天庭院设置了通往二层观景平台的楼梯，把剧场空间延伸到二层，方便容纳更多居民的参与。

空间演变

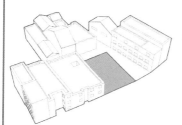

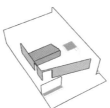

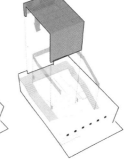

效果图

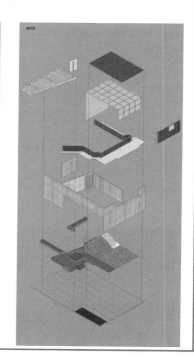

第三幕：荟所空间 ‖ The Third Act: The Living Space

荟所指的是聚集及住所。建筑处于两个传统竹筒屋的缝隙之间，面向泮塘五约主街，是一个具有办公＋住宅功能的建筑。建筑的外立面是一面高八米的镂空砖墙，光可以透过砖缝进入建筑内部，光斑洒到前庭院的墙壁上，使建筑内部有充分的光感体验。从临街的大木质推拉门进入到建筑的前庭院，仿佛置身室外，屋顶的设计采用镂空的形式，尽可能把室外的阳光引进室内。建筑一层作为穿行空间对当地居民开放，成为连接主街到内街的开口。中庭的设计为小面宽、大纵深的天井空间，改善了采光，一棵树向室内生长，将自然渗透进室内空间。室内与室外的界限在这里变得模糊，它成为人与社会与自然之间平衡关系的一种诠释。建筑分为办公、居住以及共享空间三个部分，既保证了共享空间的开放性，又保证了居住空间的私秘性。

答辯展示·民居建築保護與更新設計

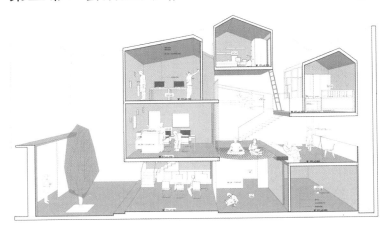

空间演变

1　　2　　3　　4　　5　　6　　7　　8

日照分析

空间分析

传统模式

现在

剖视图

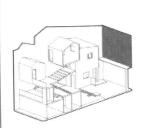 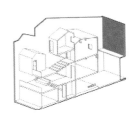 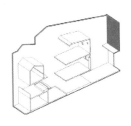

平面图

第四幕：集墟空间 ‖ The Fourth Act: The Market Space

集墟空间设计目的在于提供居民一个延伸生活的场所，预先考量其所引发和容纳的活动与用途，但也须保留无法预测的部分，以适应居民在空间中有意无意产生的互动效果。集墟空间以2米×2米×2.5米的"凹"字型钢架和布的组合形成一个装置辐射到整个空间，每个装置之间通过蜿蜒曲折的水泥板小道连接，增加集墟游玩的趣味性。

集墟空间可以根据不同时间段来灵活变动场所的功能，钢架装置的用途可能是一个临时搭建的摊位，提供给当地居民进行售卖商品，也可能通过变动布的形式形成一张休闲躺椅，供给居民一个舒适的休憩空间。另外，钢架装置和布的组合还可以形成一个小型的投影厅，晚上可以在露天进行一些放映活动。集墟还增设了一些休息座椅，可以使它变成一个小型的交流空间，供人休息交谈，满足行人的需求。在集墟的中心位置增加了一个趣味空间，方便陪同买菜的小孩在装置上游玩憩息。

空间演变

集墟装置分析

不同使用时段分析

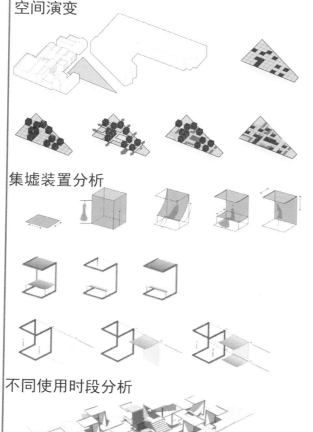

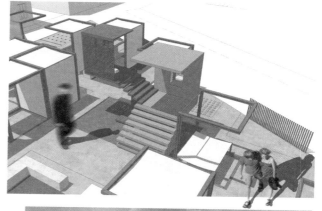

效果图

导师点评

该设计作品想法独特，采用了"集墟空间"这一设计理念，为居民提供了一系列延伸的生活空间和场所。可以看出设计前期，针对居民的需求、场地周边环境等都进行了较为深入和全面的调研及理性的分析。

设计通过研究，选择了若干节点，在不破坏原有传统建筑的基础上，利用一些装置等手法，增设了新的空间体块，使建筑之间相互联系，也为这个社区植入了更加丰富的活动功能，如剧场、集市等，形成了多层次的社区公共空间。

设计对建筑极其空间的推敲较为细致，并针对各种受众人群，不同的使用时段，都有仔细的考量。但有一点值得思考的是，位于追忆空间的天井设计，虽然很好地解决了采光问题，但从岭南地区的气候考虑，炎热的时节较长，天井在如此气候条件下的实用性还有待深入考虑。

刘晓军

A unique idea is embodied in this design work, where the residents are supplied with a series of extended living space and places with the application of design philosophy of "space of market". It can be seen that in-depth and comprehensive research and rational analysis was conducted on the needs of residents and surrounding environment during the early stage of design.

The design chooses several nodes through research, allowing interconnection among architectures through adding new space blocks on the strength of certain device without destroying the original traditional architecture. It has also added more functions for activities (including theater and market etc.), thus creating a public space of community with a multi-layered structure.

Careful considerations have been given to the design not only for the architecture and its space, but also for various group of clients at different times while using. However, when it comes to the design of patio located in the Space of Recollection, it is worth noting that there is a need for taking the practicability of such patio into account under such a climatic condition of Lingnan region featuring a long scorching season, though the lighting problem is solved.

Liu Xiaojun

专家点评

这是一件出色的毕业设计作品。之所以给予这样的评价，是从同学的作品及演讲表述中看到、听到创作小组对老城文化的理解、对老城现存问题的焦虑及设计小组对于改造老城的情怀。

通过植入青创社区，利用边角空地，设计小组试图营造新的社区交互模式，使老街在保持文脉的基础上新旧杂陈、恢复生机。另外，方案具有较强的可读性和可行性，尤其是设计小组务实学风及对当代设计语言的追求更是值得点赞。当然，评委中善意提出的在改造中应适合华南气候特点等问题需要同学百尺竿头更上一层。

王炜民

This is an excellent graduation design work. I make such a comment because I can see and hear the team's understanding of the old city culture from the design work of these students, their concern about the existing problems of the old city and their affection for the reconstruction of the old city.

The design team (innovative group) attempts to create a new community interaction model through the insertion of a youth-innovative community by utilization of corner space, allowing vigorous recovery of the old street where both the new and the old exist on the basis of the maintenance of the cultural vein of the old street. Furthermore, the design scheme is readable and feasible. In particular, it is worth giving the design team the thumbs-up for their pragmatic style of learning and pursuit of contemporary design language.

Of course, further improvements are needed by the students in terms of solution to the problems proposed by the judges, including meeting the requirement of climate characteristics in South China during the reconstruction process.

Wang Weimin

市井里院——哈尔滨市道外区中华巴洛克传统民居街区改造设计

CIID"室内设计 6+1"2017（第五届）校企联合毕业设计
CIID"Interior Design 6+1"2017(Fifth Session)University-Enterprise Cooperative Graduation Design

高　　校：	哈尔滨工业大学
College：	Harbin Institute of Technology
学　　生：	胡一非　单昳　于璐　曹在笑
Students：	Hu Yifei　Shan Yi　Yu Lu　Cao Zaixiao
指导教师：	周立军　兆翚　马辉
Instructors：	Zhou Lijun　Zhao Hui　Ma Hui
参赛成绩：	民居建筑保护与更新设计组二等奖
Achievement：	Second Prize for Folk House Protection and Update Design

胡一非　　　　　单昳　　　　　曹在笑　　　　　于璐
Hu Yifei　　　　Shan Yi　　　　Cao Zaixiao　　　Yu Lu

学生感悟

　　在周立军老师、兆翚老师及马辉老师的指导下，通过这一次传统民居的营造与空间改造设计，我们学会了用自下而上的角度出发去考虑设计。通过调研走访的方式了解守护在夕阳中的老道外的居住群体，在充分了解历史建筑背景和社会人文背景后做设计。从宏观的功能定位策划，到建筑单体空间改造甚至建筑构造设计细节都尝试着遵循改造的原真性，既是对传统民居的尊重，也是对传统民居的良性发展。

Students' Thoughts

　　Under the guidance of three teachers, Zhou Lijun, Zhao Hui and Ma Hui, we've learned to consider design from a bottom-up perspective through the design of building and spatial reconstruction of traditional dwellings. We learned about residents lived in Laodaowai as the guard in the sunset through investigation and interview, and did not do the design until we had a full understanding of the background of the historical architecture, society and humanities. We strived to follow the authenticity of the original architecture from micro planning for functional orientation to single building reconstruction and even detailed design for architecture construction, indicating our respect for and benign development of the traditional dwellings.

场地测绘 ‖ Site Survey

基于图像技术的场地估测为后续的更新设计提供了基础。

采访调研 ‖ Interviews

小组成员对 16 位场地上的居民和商铺的访谈，结论如图。

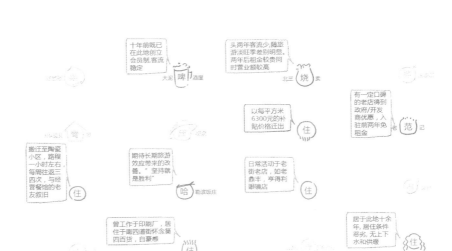

基地分析 ‖ Site Analysis

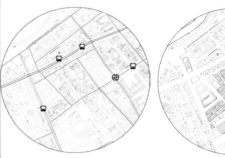

交通分析　　　　景观分析　　　　功能分析　　　　保留价值分析

餐饮休闲区

文创体验区

购物展览区

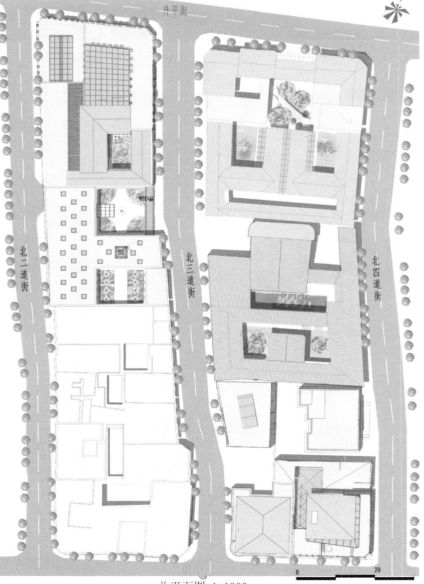

总平面图 1:1000

游客导览与自由集市区 ‖ Info Point & Free Market

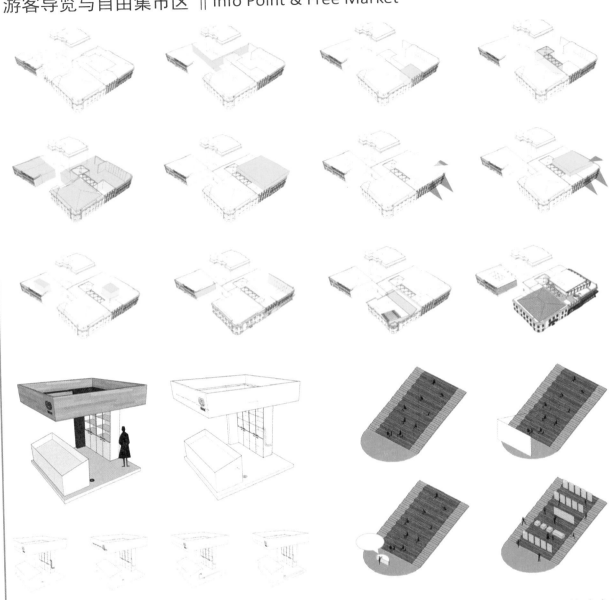

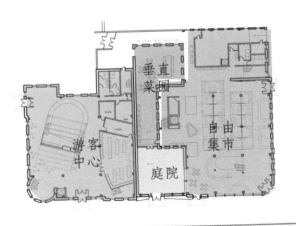

售卖单元设计，提供品牌展示及多种水电及制冷终端。

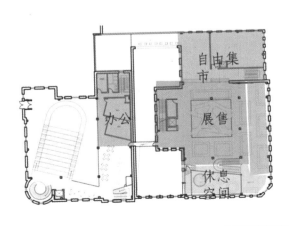

多功能大阶梯，打造复合型公共空间。

答辩展示·民居建筑保护与更新设言

文创展销与体验互动区 ∥ Handcrafts Sale & Interaction Design Zone

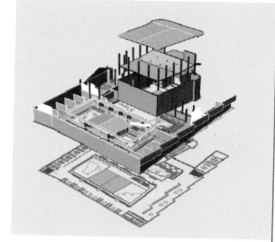

整体保留　　　　　选择性保留立面

整体拆除　　　　　拆除后在原址重建

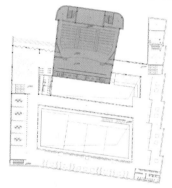

1. 尊重"中华巴洛克"工匠内涵与民族气节

老道外"中华巴洛克"传统民居作为在建筑师出现的时代前，由哈尔滨最后一批匠人打造的建筑群体，既引入了西方建筑样式潮流，同时保持了传统民居的内里，这样的时代举措不仅展示了中华民族气节，同时也是百年前匠人精神的产物与瑰宝。因此，在自由集市——活化当代业态与纪念厅——展示时代精神的两个地块之间，拟定为道外传统手工艺人展销与互动体验区，借以串联游客流线，同时保留这片区域的精神与灵魂。

2. 建筑空间处理手法

充分利用原里院空间内的临时用房，延续其意义与作用，作为新公共建筑的辅助用房——展示型工坊，定期更换工坊主题吸引游客参与并刺激消费。

餐饮休闲区 ‖ Delicacies & Leisure Zone

整体保留　　　　　选择性保留立面和山墙

整体拆除　　　　　拆除后在原址重建

3. 降低消费门槛，提高参与度

通过置入零散的消费项目，来降低消费门槛，吸引更多的本地居民和外来游客参与到道外市井文化的建设中来。

4. 打造全新的院落空间

现有改造工程所塑造的院落空间还原了"中华巴洛克"建筑的外貌，但沿用了窄巷、小门洞等原先居住模式下的院落空间，并不契合改造后的商业功能，导致在其中经营的商家无法达到预期的盈利。我们提出的解决方案是通过更改部分界面，使得院落空间的通透性和流动性得到提升，并且通过设计院落的特色景观，使得每个院落有自己的个性，打破原先均质、无趣的僵局。

5. 重建青年一代的道外记忆

由于老道外旧的商业模式与年轻人的消费观念存在分歧，现有的业态缺乏对年轻人的吸引力。长此以往，年轻人会逐渐遗忘这个街区，因为他们没有和这个街区产生互动。

通过置入美食环廊、桌游茶饮区、休闲网咖、青年旅社等适合青年人口味的新兴消费项目，来促进青年一代与这个老街区的互动，重建他们的道外记忆。

纪念厅和社区中心 ‖ Memorial Hall and Community Center

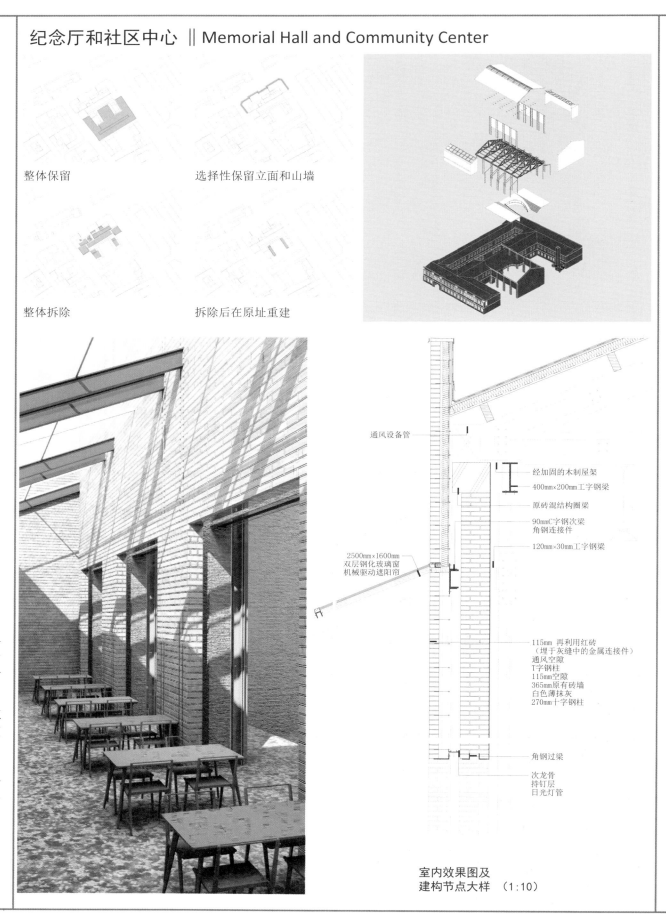

整体保留

选择性保留立面和山墙

整体拆除

拆除后在原址重建

通风设备管

经加固的木制屋架
400mm×200mm工字钢梁
原砖混结构圈梁
90mmC字钢次梁
角钢连接件
120mm×30mm工字钢梁

2500mm×1600mm
双层钢化玻璃窗
机械驱动遮阳帘

115mm 再利用红砖
（埋于灰缝中的金属连接件）
通风空隙
T字钢柱
115mm空隙
365mm原有砖墙
白色薄抹灰
270mm十字钢柱

角钢过梁
次龙骨
持钉层
日光灯管

室内效果图及
建构节点大样　（1:10）

國匠承啓卷——傳統民居保護性利用設計

导师点评

城市历史建筑保护和更新是中国城市化发展中备受关注的课题，哈尔滨工业大学同学的选题正是针对具有百年历史的老道外旧街区进行探索性的更新改造设计研究。设计成果在分析、研究、问题解决、设计表达等方面均表现出较强的逻辑性和综合设计能力，设计方案较好地将传统与现代进行有机结合，将新业态与传统业态结合，赋予街区新的活力，合理地梳理了原有旧建筑空间、创造性地营造了能够激发老街区活力的场所环境，为旧街区的复兴与可持续建设发展提出有价值的方案与构想。设计成果表现出同学们扎实的专业基本功，以及系统性的设计研究能力。如果进一步深化设计方案，希望在满足功能要求的设计方面，更细致贴切和更有创意感。

吕勤智

Urban historical building preservation and renovation, a hot topic attracting much attention in the development of urbanization in China, is selected by the students from Harbin Institute of Technology (HIT) for their design for explorative design research on the renovation and reconstruction of the ancient blocks in Old Daowai District with a history of one hundred years. A strong logicality and integrated design capability are presented in the design work in aspects of analysis, research, problem solving, design expression and the like. The tradition and modern elements are integrated perfectly in this design proposal, where there is an integration between new type of operation and traditional operation, giving new vitality to the block, making an arrangement for the original old architecture space in a proper way, innovatively creating an environment for stimulating the vitality in the old block, and supplying a valuable program and conception for the renaissance and sustainable development of such old block. Solid professional skills of the HIT students as well as their abilities for systematic design and research are demonstrated in their design achievements. For a better design proposal, I hope that the students could give more consideration to details, relevance and innovation in addition to meeting the functional requirements of the design.

Lv Qinzhi

专家点评

北三市场是哈尔滨道外区仅存的"活态"的近现代中华巴洛克历史街区，具有浓郁"滨江尘嚣"的市井文化，方案以空间尺度伦理为切入点，通过对传统空间模式的分析，在尊重历史内涵的前提下，通过小规模功能的植入，以及现代装置的加入，探讨了历史空间形态与现代功能需要的继承与发展。

朱海玄

The Beisan Market is the only "living" modern Chinese Baroque historic block in Daowai District, Harbin, where there is citizen culture with the strong flavor of "hustling and bustling Binjiang". The scheme unfolds with the Ethics of Scales, touching upon the inheritance and development of the historical space form and modern functional requirements by embedding minor functions and adding modernized equipment on the premise of respecting the historical connotation based on an analysis of the traditional space pattern

Zhu Haixuan

半山村山野 旅宿与竹韵 茶楼设计

CIID "室内设计 6+1" 2017（第五届）校企联合毕业设计
CIID "Interior Design 6+1" 2017(Fifth Session)University-Enterprise Cooperative Graduation Design

高　　校：	浙江工业大学
College：	Zhejiang University of Technology
学　　生：	高煊　周睿
Students：	Gao Xuan　Zhou Rui
指导教师：	吕勤智　宋扬
Instructors：	Lv Qinzhi　Song Yang
参赛成绩：	民居建筑保护与更新设计组二等奖
Achievement：	Second Prize for Folk House Protection and Update Design

高　煊　　　　周　睿
Gao Xuan　　　Zhou Rui

学生感悟

此次参加 "6+1" 的命题对于我们四年的本科学习来说是一次全新的挑战，很荣幸能够在毕业前接触到关于传统民居的课题。在老师的耐心指导下，我们从对课题的生疏和无从下手，到慢慢学会分析和思考，到后面成果的一步一步完成，对于我们来说这个过程无疑是艰难的，我们怀疑和否定过自己，也总是陷入难题，最后能完成这项课题研究，我们衷心地感谢指导老师和前辈们的帮助与监督。尽管我们仍存在许多问题和不足，仍然浅薄和稚嫩，但在这一过程中学习到的东西，已经比结果本身重要得太多。

Students' Thoughts

We are honored to be invited to take part in CIID "Interior Design 6+1" 2017 (the 5th session) University-Enterprise Cooperative Graduation Design and learn from peers from other universities by exchanging views. We raised questions through early systematic investigation and analysis on current status of the designed plot, then came up with the concept and finally designed and completed a project, through which our abilities were improved in various aspects. Furthermore, the reconstruction and protection of ancient villages has been a hot issue attracting much attention today. Through planning and reconstruction design for Banshan Village, we were involved in this field in a more systematic way. We realized that design shall be based on people at the very beginning of the design process; the architectural style depends on the architectural content. We had mixed feelings and gained a lot from the implementation of the graduation project.

前言 ‖ Preface

　　半山村不仅有丰富种类的游玩项目，而且其远离市嚣、竹梨环绕的自然环境，也吸引着向往归野山居、体验质朴生活的人们来到半山驻留。半山茶楼和旅宿是对外展示的窗口之一，半山村的历史文化积淀和几代村民的生活方式，都凝聚其中。

建筑空间设计 ‖ Architectural Space Design Concepts

 茶亭　面积小，路边停歇

茶肆　面积中等，有服务人员和茶点

茶楼　面积大，服务设施完善

答辯展示·民居建築保護與更新設計

茶肆设计 || Tea Shop Design

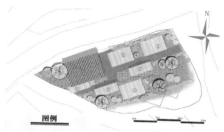

图例
① 主体空间
② 客座空间

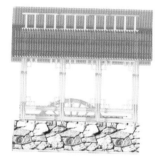

茶肆的设计以当地的竹子、蛮石为主，再用木材辅助。整个茶肆植物环绕，贴近自然。在布局上依据村庄的聚落形态错落分布，造型上既保有传统的建筑结构，又具有现代气息。

茶楼设计理念 || Design Concepts

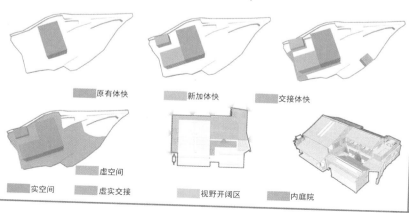

原有体块　　新加体块　　交接体块
实空间　　虚空间　　虚实交接　　视野开阔区　　内庭院

茶楼建筑设计 ‖ Restaurant Architectural Design

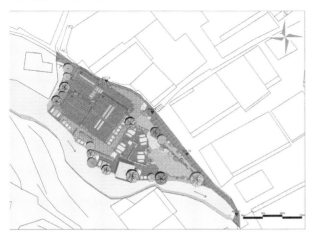

茶楼主体建筑以保护性修复为主，结合实用功能需求，在建筑周围建立亭台、廊架，供游客歇脚和饮茶。室内空间运用传统材料和现代元素，为游客营造安静舒适的茶饮空间。

答辩展示·民居建築保護與更新設計

茶楼室内设计 ‖ Restaurant Interior Design

旅宿设计理念 ‖ Design Concept of the Inn

保护　　　　改造　　　　更新

半山旅宿设计地块内的建筑有传统的"旧"建筑，也有后建的"新"建筑，作为留宿旅客的空间，对不同风貌的建筑采取保护、改造、更新三种模式，使其既传承与体现半山村村民传统生活环境与生活方式，也成为适应现代生活习惯的宜居空间。

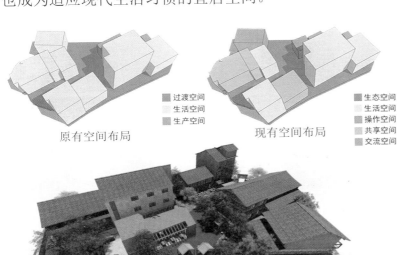

原有空间布局　　　　　现有空间布局

■ 过渡空间　　■ 生态空间
□ 生活空间　　□ 生活空间
■ 生产空间　　■ 操作空间
　　　　　　　■ 共享空间
　　　　　　　■ 交流空间

答辩展示·民居建築保護與更新設計

半山旅宿接待处设计 ‖ Reception Design

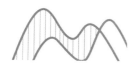

建筑外形为风貌保存较好的传统木质民居，是半山村文化的遗存，也是半山村的名片之一，将其作为半山旅宿的接待处，为来到半山旅宿的游客提供入住登记、行李寄存等服务。

半山旅宿公共餐厅设计 ‖ Public Restaurant Design

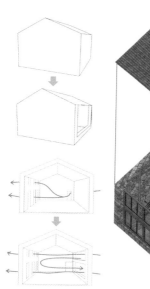 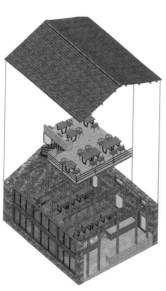

拆除原有建筑的部分墙体，使用大面积可折叠推拉的落地玻璃窗，增加室内采光与通风，将庭院空间引入室内，庭院空间与餐厅空间失去了界线，旅客们仿若在自然环境里用餐和聚会，增加旅客们对山村自然环境的感知。

半山旅宿咖啡厅设计 ‖ Cafe Design

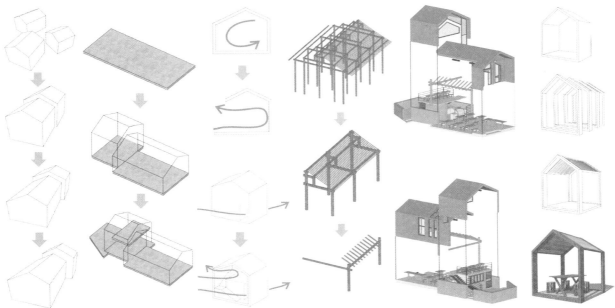

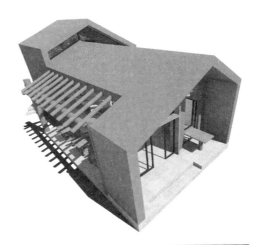

以半山村建筑布局为启发,将不同体量的建筑体块叠加组合,再对体块进行推移和切割,前部分作为咖啡厅的座位区,后部分一层室内作为操作区,室外作为户外烧烤区,二层作为户外烧烤区的用餐区。前部分体块一部分为半露天座位,结合半山村传统民居建筑,运用穿斗式和抬梁式结构,简化传统的形式语言,将其作为建筑的构件。

旅宿家庭房型民宿设计 ‖ Residence Design

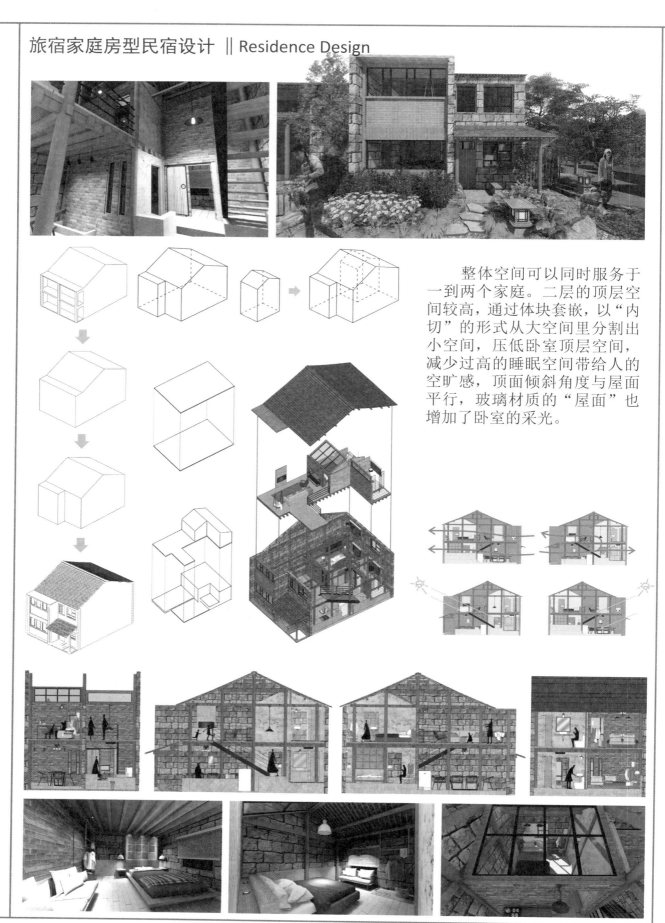

整体空间可以同时服务于一到两个家庭。二层的顶层空间较高，通过体块套嵌，以"内切"的形式从大空间里分割出小空间，压低卧室顶层空间，减少过高的睡眠空间带给人的空旷感，顶面倾斜角度与屋面平行，玻璃材质的"屋面"也增加了卧室的采光。

导师点评

浙江省富山乡半山村传统民居保护与利用设计,学生通过半山村田野考察、实地勘测,对场地、水文、植被做了普查和综合分析,对人口分布状况、地域特色文化、全域旅游资源、目标人群定位等做了系统研究、在保护村庄肌理、传统民居风貌前提下进行了环境整治、景观设计、茶亭、茶肆小品等设计,对建筑进行改造利用,丰富休闲功能,做了民宿、茶楼、咖啡厅等空间环境设计,能够掌握设计程序、控制设计手法、设计深度满足毕业设计要求,可实施性强,如能再简约、淳朴些,乡土气息会更自然。

杨琳

This design work relates to the protection and development of traditional dwellings in Banshan Village, Fushan Township, Zhejiang Province. Through the field investigation and survey in Banshan Village, and general investigation and comprehensive analysis on the site, hydrology and vegetation, the students conducted systematic research on population distribution, distinct local culture, the overall tourism resources and the identification of target population. They conducted environmental remediation and made design in the landscape, tea pavilion and teahouse sketch on the premise of protection of village texture and style and features of the traditional dwellings, for the purpose of reconstructing and utilizing the architecture to add more functions for leisure propose, and make environment design in space, including the home stay room, teahouse and café. The students are able to master the design process, control the design method and depth so as to meet the requirement for graduation design with good feasibility. Their design work will be more natural if brief, simple elements and local flavor are incorporated.

Yang Lin

专家点评

该项目对于村落的前期调研能够作出系统的分析,整体设计具有一定的完整性,能抓住主要元素做组合式研究。

但好的设计是击中痛点并有故事性。如果设计的图纸效果表现增加些张力和想象力,会产生感染力并增添项目魅力。题目是对富山乡半山村对传统民居保护与开发,但对民居的特征及保护价值分析应该更深入,设计应深挖其地域性,再加以适当地创新。项目同时应增加对空间与其周边环境、业态的分析,空间设计的本质是对时间的设计,空间在一天、一周、一年四季怎么用都是有差异和不同价值的。

姚领

Systematic analysis is made in the project concerning the early investigation on the village. The overall design is integrity to some degree, where modular research is made through grasp of main elements.

However, good design is able to hit the key point with interesting plot. Appeal and charm of the project will be created if more tension and imagination is embedded into the effect performance of the design drawing. The design title refers to protection and development of traditional dwellings in Banshan Village, Fushan Township, Zhejiang Province, so a deeper analysis on the characteristics and protection value of the dwellings shall be made. The design shall pay attention to the locality and contemporary innovation. Moreover, analysis on the space, its surrounding environment and type of operation shall be added into the project. Space design is, in essence, the design of time, for space in one day, one week or different seasons of a year are varied with different values.

Yao Ling

上海近代黑石公寓及其保护周边环境保护计与更新设计

CIID"室内设计 6+1"2017（第五届）校企联合毕业设计
CIID "Interior Design 6+1" 2017 (Fifth Session) University-Enterprise Cooperative Graduation Design

高　　校：	同济大学
College:	Tongji University
学　　生：	李淑一　常馨之　安麟奎
Students:	Li Shuyi　Chang Xinzhi　An Linkui
指导教师：	左琰　黄全
Instructors:	Zuo Yan　Huang Quan
参赛成绩：	民居建筑保护与更新设计组三等奖
Achievement:	Third Prize for Design and Renovation of Residential Buildings

李淑一
Li Shuyi

常馨之
Chang Xinzhi

安麟奎
An Linkui

学生感悟

随着毕业日子的到来，毕业设计也接近了尾声。经过几周的奋战我们的毕业设计已经完成。在此要感谢左琰老师对我们悉心的指导，感谢组员之间互相帮助以及共同面对压力的互相安慰和寻找动力的互相鼓励。在设计过程中，我们通过查阅大量有关资料，彼此的交流经验和自学，并向老师请教等方式，学到了不少知识，也经历了不少艰辛，但收获巨大。通过毕业设计，我们懂得了许多东西，也培养了独立工作的能力和动手的能力，相信会对今后的学习工作生活有非常重要的影响。

Students' Thoughts

As the time for graduation is approaching, the deadline for graduation project completion is also coming. We've finished our project following several weeks' efforts. We hereby would like to express our appreciation to our mentor Zuo Yan for his careful guidance, and to the team members for their help, mutual consoling when facing pressure and mutual encouragement for seeking impetus. During the implementation of the project, we exchanged experience and conducted independent study by reference to plenty of pertinent information, asked for suggestions from the mentor, learned things and experienced hardships, but we also harvested a lot. We have acquired a lot and developed independent working ability and operational skills, which we believe will indeed have a major impact on our further study, work and daily life.

设计概念生成 || Concept Formation

这里是曾经的法租界,海派文化盛行,历史建筑集中。面对基地封闭现状,我们引入"聚落"的概念,开放周边入口,建立步道系统,通过人群、业态、空间场景的混合,重现基地的历史。同时将街区特色——"音乐"元素加入设计中,赋予基地当代特色。

1. 封闭的基地现状

2. 拆除加建部分 打开围墙

4. 地下停车 人车分流

3. 恢复花园 建立步道系统

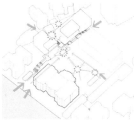

5. 多入口 多节点 多样人群

6. 植入观景展示体块 加强街区互动

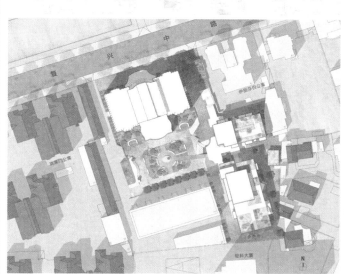

总平面图

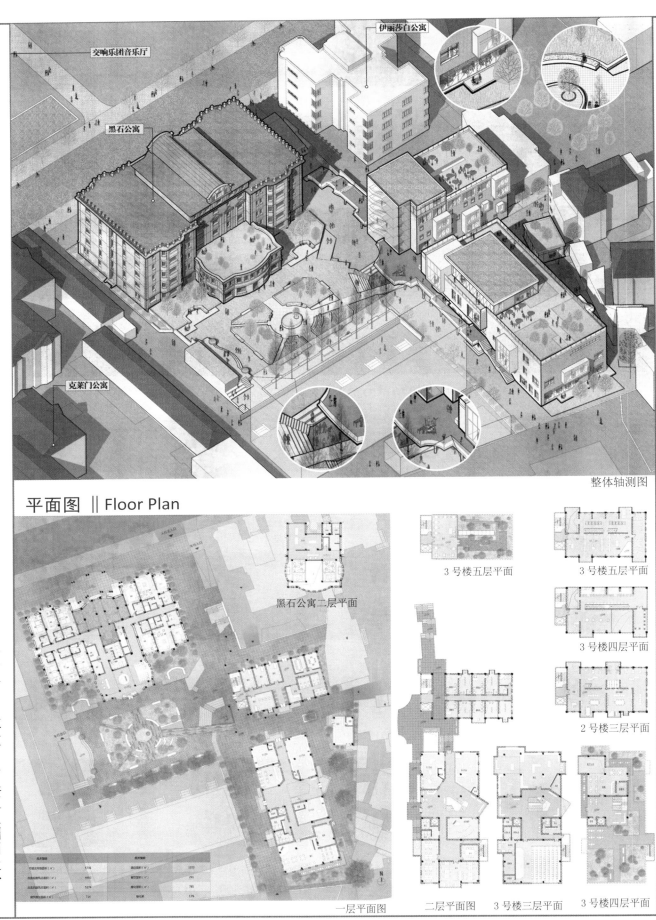

场地设计 ‖ Site Design

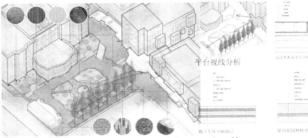
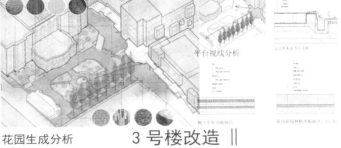

花园生成分析

3号楼改造 ‖ Building No. 3 Reconstruction

1 原有轴线生成古典对称平面

2 根据空间节点，生成新轴线

3 新的人行系统形成

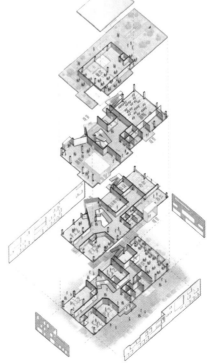

答辩展示·民居建築保護與更新設計

历史记忆拼贴——2号楼设计 || Historical Memory Collage--Building No.2 Design

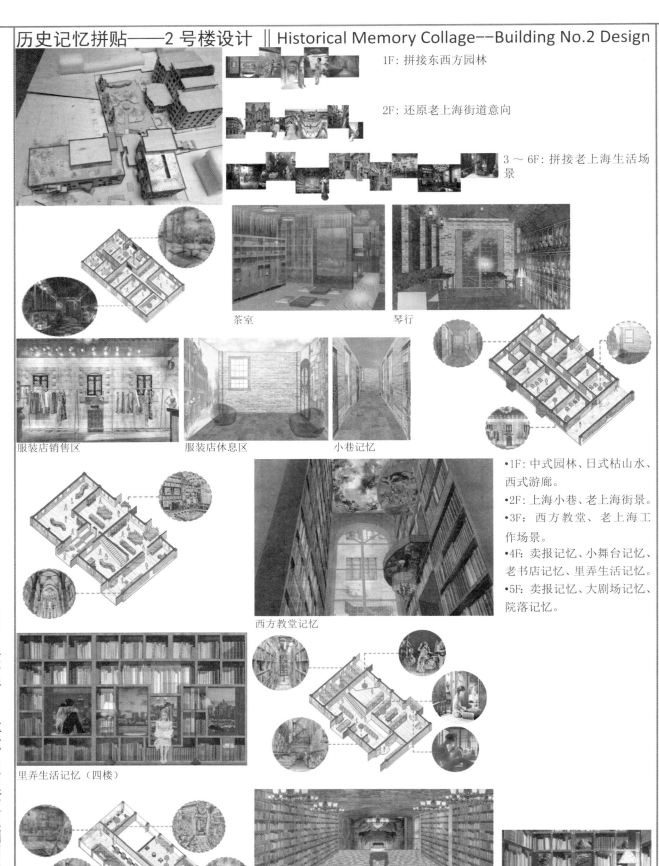

1F：拼接东西方园林

2F：还原老上海街道意向

3～6F：拼接老上海生活场景

茶室　　琴行

服装店销售区　　服装店休息区　　小巷记忆

- 1F：中式园林、日式枯山水、西式游廊。
- 2F：上海小巷、老上海街景。
- 3F：西方教堂、老上海工作场景。
- 4F：卖报记忆、小舞台记忆、老书店记忆、里弄生活记忆。
- 5F：卖报记忆、大剧场记忆、院落记忆。

西方教堂记忆

里弄生活记忆（四楼）

大剧场记忆　　里弄生活记忆（五楼）

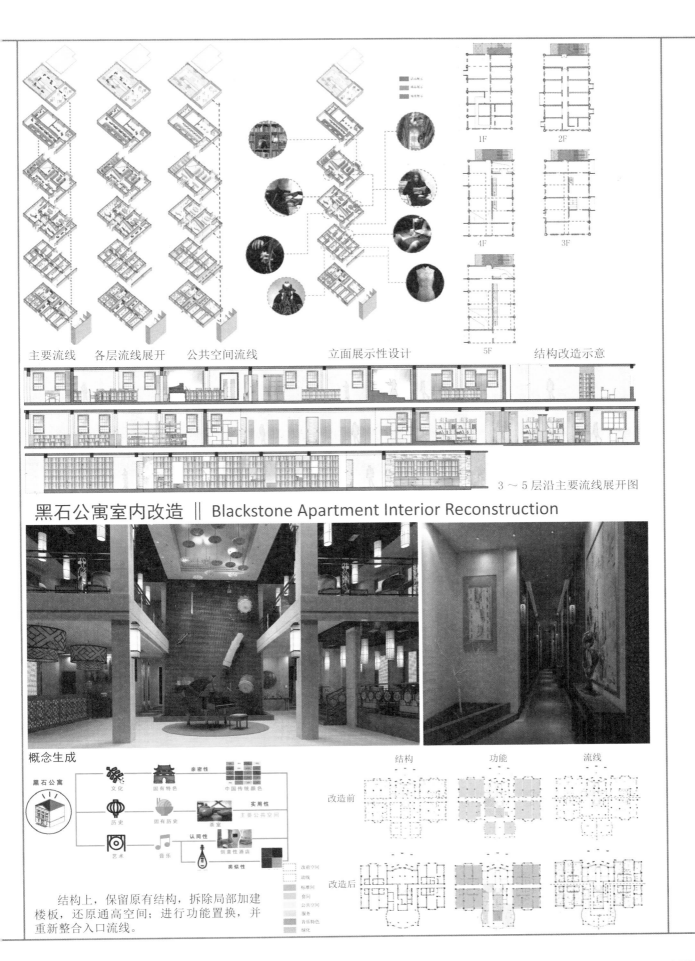

酒店客房类型 A, B ‖ Type-A, B Hotel Room

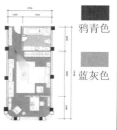 鸦青色 / 蓝灰色  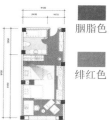 胭脂色 / 绯红色

各个房间使用不同的中国传统配色，为客人带来不同的体验。

酒店客房类型 C ‖ Type-C Hotel Room

 茶色 / 牙色

酒店空间构成 ‖ Component of Hotel Space

材质　　　顶面　　　公共空间

松木
黄洞石
剑麻地毯
白棕色大理石
石砾
瓷砖

野樱桃木
古木
松木
山毛榉木
黄洞石
剑麻地毯
深蓝色地毯

剖面图 ‖ Section

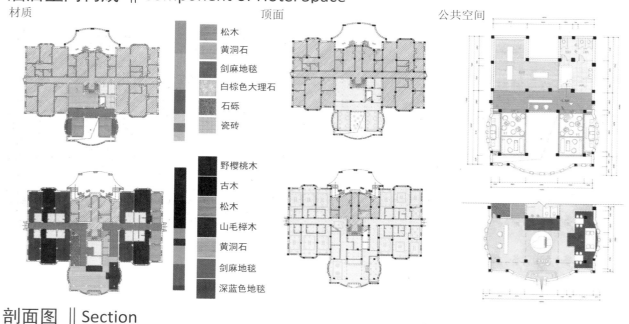

A-A'　　　　　　　　　　　　　　　　　　　　　　　　　　C-C'

导师点评

　　项目设计条件较为复杂，现状用地狭窄，车流、人流混杂，经过长时间的使用，历史建筑和新建、加建建筑不协调，改造设计的难度较大。

　　该设计合理地利用了高差，通过下沉广场、挑高的平台，建立了多层次的步道系统，解决了交通问题，同时也丰富了场地的外部空间，并在几个新建建筑之间建立了联系。在黑石公寓的后面恢复了民国风格的的花园，延续了历史文脉。功能方面，考虑了酒店、餐厅、音乐教室、手工艺制作、服装店、书店、茶室、展示、办公、剧场等多样的使用空间。室内设计部分，用乐器作为装饰元素，显示了社区和音乐方面的历史渊源，又加入了中式的装修风格，并考虑了各方面的建筑材质。图面表达清晰，排版紧凑，内容丰富，显示了一定的绘图功底。和周边建筑间距较近，改造时应该适当考虑防火方面的要求，个别楼梯画法有误，个别房间没有标出使用功能。

石拓

　　Project design conditions are relatively complex. Reconstruction design is a difficult task as the newly-constructed or added buildings fail to match with the historical buildings following long-term use and the pedestrians are mixed with the vehicles disorderly on the narrow land used in current status.

　　Traffic problem is resolved in the design by establishment of a multi-layer footpath system with sunken plaza and overpass platform based on height difference. At the same time, connections are made among several newly-constructed buildings as a result of the broadening outer space. History heritage is preserved as the ROC-style garden is restored at the back of the Blackstone Apartment. Considerations are given to functional application, such as hotel, restaurant, music classroom, handicraft production, clothing store, bookstore, tea room, display, office, theater and other various application of space. In terms of interior design, musical instruments are used as the decorative elements to demonstrate the community's historical origin in terms of music. In addition, Chinese decoration style is added in the design with consideration of a wide range of building materials. Good drawing skills are shown by the clear drawing expression, compact layout and the enrichment of the content.

Shi Tuo

专家点评

　　一个城市的活力往往取决于对老建筑和历史文化的包容度以及理性的设计，同济2组同学的设计很好地延续了海派文化的包容性和多元化的特征，将黑石公寓特有的建筑语汇和时代特征和旁边不同时期的建筑做了相应的融合，用更符合当下审美和功能的手法，与历史建筑发生对话，使原本平淡的建筑重新焕发了生机。功能上也能从区域特征着手，从新融入新的业态，让整个建筑群落更为丰满。室内设计更多是想留住过去的历史，所以更多的是还原过去，如果能有更为时尚和前卫的手法植入，会给整个设计带来更耳目一新的感受。

黄全

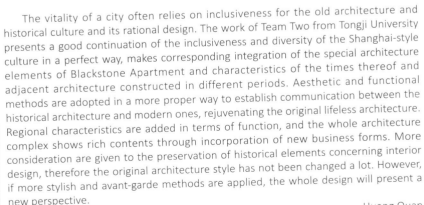

　　The vitality of a city often relies on inclusiveness for the old architecture and historical culture and its rational design. The work of Team Two from Tongji University presents a good continuation of the inclusiveness and diversity of the Shanghai-style culture in a perfect way, makes corresponding integration of the special architecture elements of Blackstone Apartment and characteristics of the times thereof and adjacent architecture constructed in different periods. Aesthetic and functional methods are adopted in a more proper way to establish communication between the historical architecture and modern ones, rejuvenating the original lifeless architecture. Regional characteristics are added in terms of function, and the whole architecture complex shows rich contents through incorporation of new business forms. More consideration are given to the preservation of historical elements concerning interior design, therefore the original architecture style has not been changed a lot. However, if more stylish and avant-garde methods are applied, the whole design will present a new perspective.

Huang Quan

答辯展示·民居建築保護與更新設計

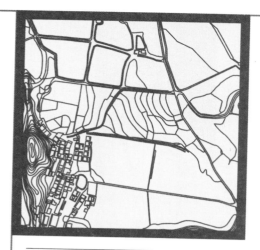

一房一社区

CIID"室内设计 6+1"2017（第五届）校企联合毕业设计
CIID"Interior Design 6+1"2017(Fifth Session)University-Enterprise Cooperative Graduation Design

高　　校：	华南理工大学
College：	South China University of Technology
学　　生：	林嘉辉　陈小帆　卢惠兴　张裕麟
Students：	Lin Jiahui　Chen Xiaofan　Lu Huixing　Zhang Yulin
指导教师：	薛颖　石拓　骆雯　谢冠一
Instructors：	Xue Ying　Shi Tuo　Luo Wen　Xie Guanyi
参赛成绩：	传统民居保护与更新设计三等奖
Achievement：	Third Prize for Folk House Protection and Update Design

林嘉辉
Lin Jiahui

张裕麟
Zhang Yulin

陈小帆
Chen Xiaofan

卢惠兴
Lu Huixing

学生感悟

　　短短三个月的毕业设计不仅仅是一次专业上的探索，更是一次文化保护与传承之旅。而"6+1"校企联合毕业设计除了让我们收获专业上的探索经验，更让我们得以深入了解传统广州本土特色骑楼，真正地去思考如何让曾经凝聚民智的民居老建筑得以焕发新生。从前期调研、概念的建立到方案思路的推导、功能分析，种种细节都凝聚了我们四个人半年的心血。我们见识了清晨6点刚刚醒来的广州老城区，了解了北京路一带传统建筑的历史，拜访了仍然致力于传承传统工艺的老匠人等。衷心感谢一直给予我们指导的各位老师，感谢他们不断为我们的方案提出建议，伴随我们在设计中成长。

Students' Thoughts

　　The graduation project completed in only three months is a professional exploration and also a journey of cultural protection and heritage. Thanks to the 6+1 graduation project initiated by the joint efforts of the universities and enterprise, we has obtained professional experience through exploration, and more importantly, we have got an in-depth understanding of traditional arcade with local characteristics in Guangzhou and seriously think about how to bring new vitality to the traditional dwellings embodying wisdom of people. Our six months' painstaking efforts have been put in particulars from early survey, establishment of concept to deduction of scheme thought and functional analysis. We have seen the old city of Guangzhou at 6:00 a.m., learnt the history of the traditional architecture in the area of Beijing Road, and paid a visit to old craftsmen who have been committed to traditional handicraft inheritance.. We would like to express our sincere appreciation to our mentors for their constant suggestion concerning our project proposal and their support in our progress during the implementation of our project.

A COMMUNITY
MORE THAN A BUILDING: 一房即一社区

概念构思 ‖ Concept Design

众

"众"这个概念可以引申很多关键词，如众事、众聊、众知、众乐、众憩等，从而形成办公空间、社交空间、知识空间、玩乐空间、休闲空间。这些延伸出来的空间也正是我们社区活动中心主要的空间属性。

从前，居民生活融洽，群体间不单是相互聚焦，而且相互联系，是真正意义上的社区。

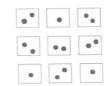

现在，随着时代发展，像上班族、留守儿童、独居老人产生，逐渐演变成仅仅居住在一起的独立集体。

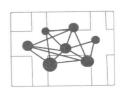

"种"是大家一起的意思，我们通过这个设计试图增加小区里各类群体的联系和交流，重新阐述社区的概念。

	众知	知识空间 KNOWLEDGE SPACE
	众乐	玩乐空间 PLAY SPACE
	众事	办公空间 OFFICE SPACE
	众联	社交空间 SOCIAL SPACE
	众憩	休闲空间 LEISURE SPACE

答辩展示·民居建築保護與更新設計

建筑空间构思 ‖ Architectural Space Design

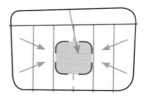

保留部分原有的建筑结构形成大的公共空间并串联其他空间

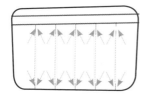

较完整地保留原有结构，促进空间交流

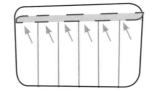

天井中绿植景观给人们带来了视觉上的享受

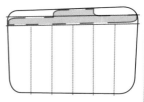

扩充天井增加使用者的公共活动空间

原有的孤立乏味，相互隔离的单开间

打破这个模式，提供更多的交流机会

功能分析 ‖ Functional Analysis

由于打开了两个开间成为中庭体块，所以空间布局围绕中庭展开。原有空间仅靠一条直通的狭长楼梯进行楼层之间的连通，缺少空间之间的交流与联系。我们试图通过镂空楼层、设置观景阳台等方法，并在建筑外增加楼梯间和电梯间，连通上下空间，增强空间连续性，丰富了视觉感受和空间体验。

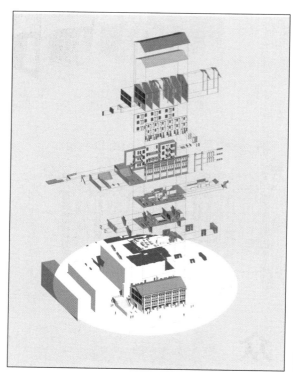

平面图 ‖ Floor Plan

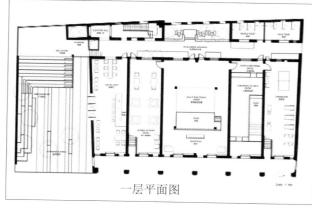

一层平面图

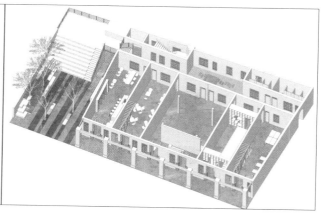

一层将部分旧墙体拆除，在中间形成大的活动空间汇聚人流。一层主要设置了咖啡厅（众憩空间）、老人活动室（众乐空间）、舞台（众乐空间）、多功能活动室（众乐空间）、儿童活动室（众乐空间）和健身室（众乐空间），以"众乐"和"众憩"空间为主。

答辯展示·民居建築保護與更新設計

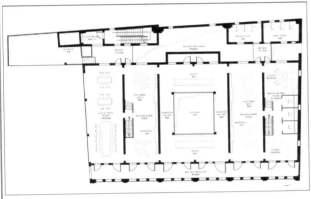

将中间部分墙体拆除，替换结构，形成一个大的镂空。促进层与层之间的视线交流，给使用者带来开阔的视觉感受。

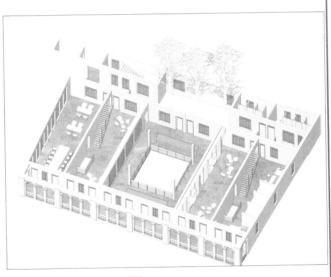

二层平面图

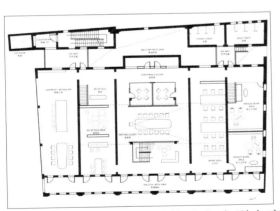

打造一个集手工互动体验和小型电商办公于一体的"众事"空间。

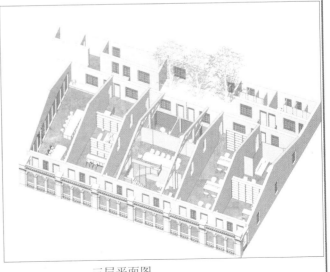

三层平面图

改造前后空间对比 || Contrast Before and After Interior Reconstruction

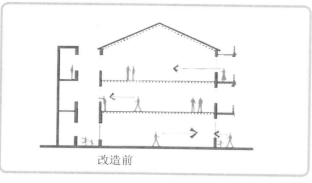

改造前

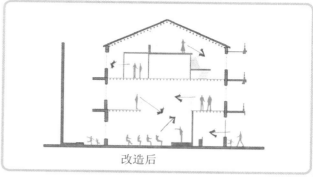

改造后

　　原先室内空间进深大，狭窄成条，依赖天井进行采光通风，因不能满足整栋建筑需求而导致室内空间潮湿阴暗，尤其底层空间。我们扩大了原先建筑赖于采光的天井，并在建筑外墙适当地开落地玻璃窗，计算全年日照角度及穿堂风流线，适当的在屋顶以及墙面开窗，改善室内空间的采光与通风条件。

　　保留建筑内部原有的楼梯，在副楼中增设了电梯间、楼梯间和平台通道，提供多条活动路线，增添空间的趣味性，使进入到空间内的人流迅速疏散。不同使用者有各自不同的流线，互相之间不会干扰，但又能在公共活动区域汇集到一起。

室内 || Indoor

将中间部分墙体拆除，替换结构，形成一个大的镂空。使二层的使用者看到一层中间的舞台，促进层与层之间的视线交流，给使用者带来开阔的视觉感受。二层主要设置了市民活动室（众乐空间）、阅览室（众知空间）、展厅（众知空间）、私人阅读区（众知空间）和景观阳台（众憩空间），"众乐""众知""众憩"空间的相互结合，营造出多元化的社区氛围。

本次设计旨在修复与新建、老料与新材、已有空间与当代设计相结合。现代住宅社区与休闲等新的功能较为自然地融入了老的较为私密与生活化的环境。活化与再利用历史建筑，并使其为周边社区居民服务。

采光与通风 ‖ Day-lighting &Ventilation

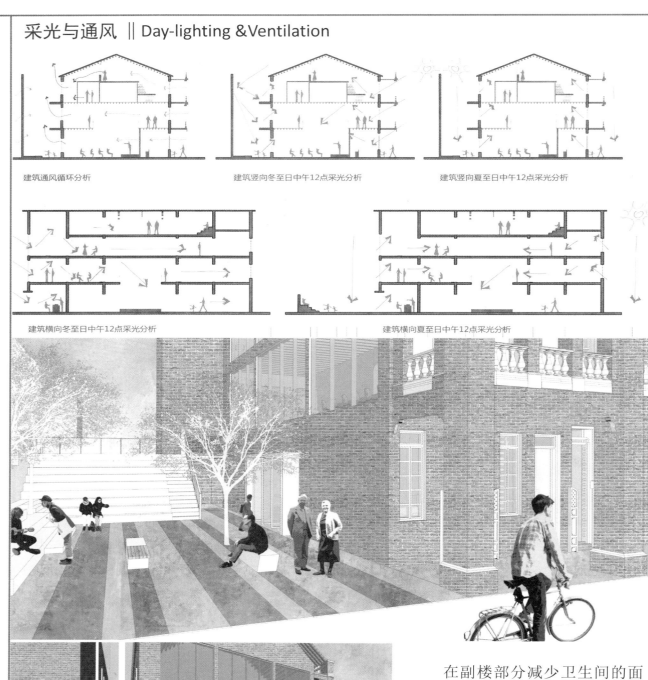

建筑通风循环分析　　建筑竖向冬至日中午12点采光分析　　建筑竖向夏至日中午12点采光分析

建筑横向冬至日中午12点采光分析　　建筑横向夏至日中午12点采光分析

在副楼部分减少卫生间的面积，增加了电梯间和楼梯间，搭建了广场通向二楼平台的楼梯，同时还扩大了部分天井的面积。建筑外的广场设置了一些大阶梯，这些作为"众憩空间"，供人们休闲和观赏风景。

导师点评

　　"一房一社区"顾名思义在这栋老房子里建立起了一座供人们共同使用的内容丰富的社区空间。

　　方案设计之初，设计者提出"众"这个概念，众事、众聊、众知、众乐、众憩等，从而产生了办公空间、知识空间、玩乐空间、休闲空间，这些延伸出来的空间也正是社区活动中心的主要空间属性。设计者对原有的狭长单开间进行了大胆改造，替换建筑中间的砖墙结构，适当打通隔墙，形成大空间，建筑原有结构完整地保留下来，通过相同的天井串联建筑前部互相独立的空间，丰富了空间的内容。

　　建筑平面布局丰富且合理，认真考虑了结构问题，并且对通风、日照进行了有效分析。设计方案内容丰富、逻辑清晰、表现力强，图纸深度到位，让原有受到限制的空间焕发新的光彩。

朱宁克

　　The title "One house is One Community" implies creating a community space in the old house with rich contents for people's common use.

　　The designer has proposed the concept of "public" since the inception of the project design, namely public issues, consecration, knowledge, entertainment and recreation, etc., giving birth to extended space, including office, knowledge, recreation and leisure space, which is the main spatial attribute of the community activity center. The designer boldly implements reconstruction of the original narrow single-bay room to replace brick wall structure in the middle of the architecture, properly knocks down the partition wall to form a larger space, completely preserves the integrity of the original architecture structure, and connect the mutually independent spaces in the front of the architecture through the same patio, thus enriching the content of the space.

　　In the dense and reasonable architectural plane layout, serious consideration is given to the structure issues, and effective analysis has been conducted on ventilation and sunlight. The design proposal is characterized by rich content, clear logic, good expressiveness and in-depth drawing presentation, thus bringing out new vitality of the original limited space.

Zhu Ningke

专家点评

　　该项目设计倡导一种"众"的开放社区模式，从冷漠走向交流，从封闭走向开放。在空间功能、动线、通风、采光等方面都能体现设计者对空间的功能探索和价值思考。但如果项目跟周边环境、业态、交通等分析做进一步深入调研，更系统地思考开放社区模式，以及其空间设计与实际使用人群的行为心理研究会更精彩。设计不能单纯理解为在"美学"和想象里做功夫，不仅是对比例、尺度、色彩的有机组合，而应该是其价值观，文明、平等、自由、生态、人文等各领域地平衡与设计。项目需要结合实际情况，深入而细腻地解决问题。

姚领

　　The project design advocates a "public" open community model, where people start to have the communication with each other without indifference, and the closed community is evolving into an open community. Functional exploration and value reflection by the design to the space are embodied in aspects of spatial function, moving line, ventilation, lighting and so on. However, if further in-depth research is conducted on the analysis of the relationship between the project and surrounding environment, business forms and traffic, and more systematic considerations are given to the model of open community, a better work will be done regarding spatial design and research on the behavior psychology of the actual users. The design shall not be simply focus on the "aesthetics" and imagination- namely the organic combination of scale, dimension and color, but on the balance and design in various fields such as value, civilization, equality, freedom, ecology, and humanities etc. The project should be linked with reality so as to resolve the problems in a deep and meticulous manner.

Yao Ling

里里·里院
哈尔滨老道外传统街区院落再利用与里院商业的探索

CIID "室内设计 6+1" 2017（第五届）校企联合毕业设计
CIID"Interior Design 6+1"2017(Fifth Session)University-Enterprise Cooperative Graduation Design

高　　校：	哈尔滨工业大学
College:	Harbin Institute of Technology
学　　生：	李焱　郭冰燕　樊逸冰　刘名淞
Students:	Li Yan　Guo Bingyan　Fan Yibing　Liu Mingsong
指导教师：	周立军　兆翚　马辉
Instructors:	Zhou Lijun　Zhao Hui　Ma Hui
参赛成绩：	传统民居保护与更新设计组三等奖
Achievement:	Third Prize for Folk House Protection and Update Design

李焱　　　　　郭冰燕　　　　樊逸冰　　　　刘名淞
Li Yan　　　 Guo Bingyan　 Fan Yibing　 Liu Mingsong

学生感悟

　　这次毕业设计是对大学五年建筑学专业学习总结，一个学期的相处时光及小组成员的默契合作使我们的毕业设计变得格外有意义。大一时我们从调研哈尔滨传统建筑开始学习建筑学，大五我们深入探讨哈尔滨传统民居的未来发展和改造设计，是对我们五年建筑学业生涯的完整收尾，也是我们对哈尔滨这座城市的深切希冀。我们在设计的过程中分享知识，交流见解，共同作业，一起成长，也收获了珍贵的毕设革命友谊。今后我们都将开始新的学习，我们会带着毕设中我们对中国传统建筑的思考继续前行。

Students' Thoughts

In addition to the independently completed summarization of our five years' study of architecture in HIT, the tacit cooperation of our team members during the semester has made our graduation project particularly meaningful. We started to learn architecture by investigation and survey on the traditional architecture in Harbin during the freshman year in HIT, and made in-depth exploration about the future development and reconstruction design of traditional dwellings in Harbin as graduates during the fifth year, which ends our five years' academic career in architecture and shows our heartfelt expectation of Harbin. During the design process, we shared knowledge, exchanged opinions, worked and made progress together, establishing a precious friendship with each other. We also learnt a lot from the teachers and students from other six universities through discussion. The time for graduation is approaching, and we are ready to start a new phase of study. We will make further progress with our thinking about traditional Chinese architecture indicated in our graduation project.

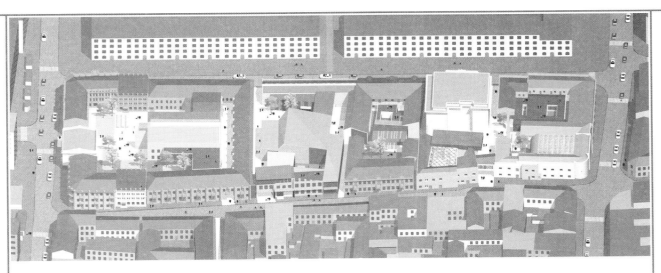

建筑空间构思 || Architectural Space Design Concepts

基地为一块完整街区

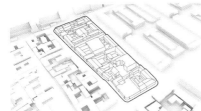
拆除不具备使用功能的建筑

连结院落开放空间增加商业界面

新加体量保持街区内原有尺度

为大空间引入天光还原院内外沟通

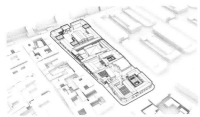
屋顶绿植平台增加街区绿化率

功能和流线 || Function and Route

新形成的内街结合原有里院机理，形成了一次分布在内街的四个核心院落区。对应四个重要的改造节点，形成了以书店、剧院、电影院和商业为主的四个功能区。

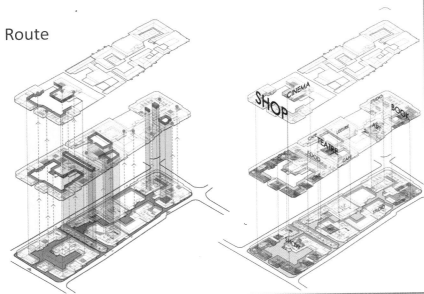

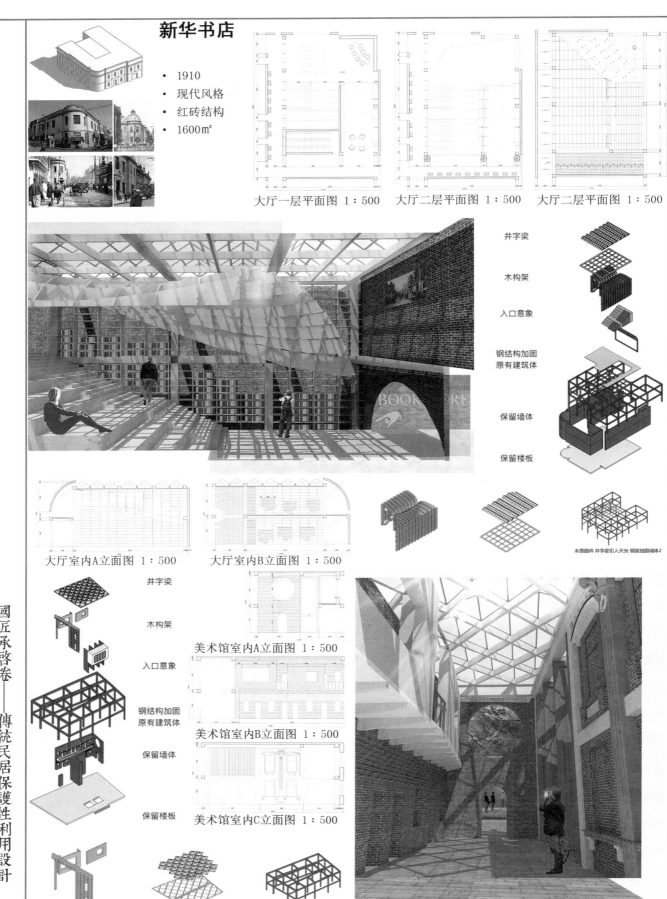

黑龙江省评剧院

- 1940
- 现代风格
- 框架结构
- 1700m²

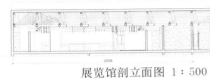

保留原屋顶
保留原屋架
加入吊挂灯具
加入展板、展架
保留原砖墙

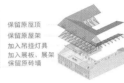

展览馆剖立面图 1:500　　展览馆剖立面图 1:500

展览馆剖立面图 1:500　　展览馆剖立面图 1:500　　展览馆剖立面图 1:500　　展览馆剖立面图 1:500

咖啡厅剖立面图 1:500　　咖啡厅剖立面图 1:500　　咖啡厅剖立面图 1:500　　咖啡厅剖立面图 1:500

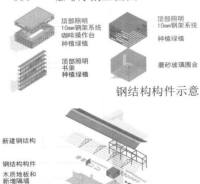

顶部照明
10mm钢架系统
咖啡操作台
种植绿植

顶部照明
10mm钢架系统
种植绿植

顶部照明
书架
种植绿植

磨砂玻璃围合

钢结构构件示意

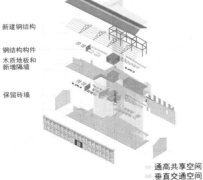

新建钢结构
钢结构构件
木质地板和新增隔墙
保留砖墙

■ 通高共享空间
■ 垂直交通空间

答辩展示·民居建筑保护与更新设计

松光电影院

- 1940
- 现代风格
- 框架结构
- 2600 ㎡

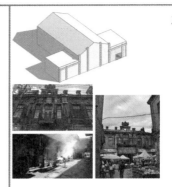
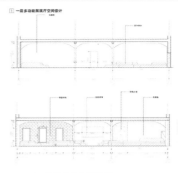

多功能厅立面图 1:500

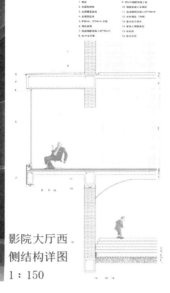

影院大厅西侧结构详图 1:150

放映厅立面图 1:500

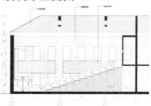

放映厅立面图 1:500

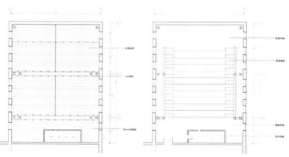

放映厅天花图 1:500　　放映厅地面铺装图 1:500

A—A剖面图 1:500

國匠承啓卷——傳統民居保護性利用設計

北三市场

- 1945
- 折衷主义
- 砖混结构
- 1800 ㎡

手绘效果图

休息室日光效果图

休息室夜间效果图

服装店地面布置图 1:500

服装店棚面布置图 1:500

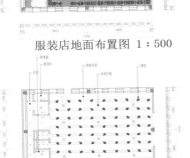

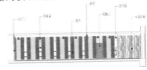

服装店墙剖立面图A 1:500

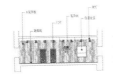

服装店墙剖立面图C 1:500

服装店墙剖立面图B 1:500

服装店墙剖立面图D 1:500

服装店夜间效果图

地面布置图 1:500

棚面布置图 1:500

墙剖立面图C 1:500

墙剖立面图A 1:500

墙剖立面图B 1:500

墙剖立面图D 1:500

短租房日光效果图

答辩展示·民居建筑保护与更新设计

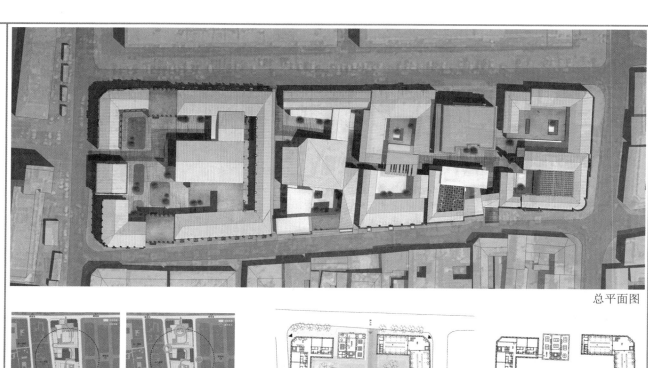

总平面图

活化界面　　事件节点

用地　　　　开放空间

里院内街，街区建筑空间的保留、依照历史建筑意象的室内空间设计、原有街区共享活动功能的延续，意在将街区的历史空间和现实改造空间形成连续的使用记忆联系。形成有传承、有特色、有思考的传统民居的改造新生。

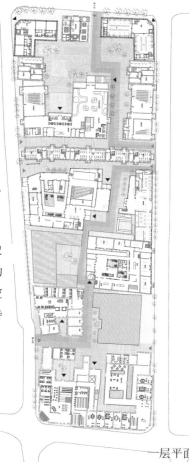

一层平面图

二层平面图

基地建筑形态

导师点评

题目"里里·里院"立意明确，点明设计主题，从三个层次诠释了传统道外里院街区的特质。利用串联的手法将四个院落联系起来，成为一个相互联系的整体，使沟通更为紧密。

值得一提的是，设计者们前期对于基地情况的分析十分细致，从不同角度、多个方面以及不同时间点，对原有建筑及环境进行了深入细致的调研。而且，对于建筑内部的构造等都做了深层次的剖析，工作细致。但是，这里要提出一点建议：首先，既然是相互串联的四个院落，它们在建筑风格上又基本属于同一时期作品，设计手法都有着较高的相似度，是否在设计每个院落的内部空间时，也能更多地考虑它们的统一性，虽然每个代表建筑的功能不同，但室内设计风格上可以整体进行考虑，目前的每个建筑室内风格无论从材质还是设计元素上都表现出较大的差异，感受不到是在一个串联的整体街区之中。其次，前期调研内容中的部分内容，如民俗和生活元素的提取，并未在后期的设计中反映出来，这是比较遗憾的地方。

总的来说，设计作品对基地功能空间的考虑，布局明确，交通流线清晰，体现了设计者扎实的基础专业知识以及分析问题的能力和水平。

古秋琳

The title "Triple Garth" has a clear conception and points out the design theme, demonstrating the characteristics of traditional garth blocks in Daowai District (Harbin) from three aspects. The interconnected architectural complex is made up of four garths through series connection, allowing a closer communication among them.

It is worth mentioning that, during the early stage, detailed analysis of base condition was made by the designers, who made in-depth and meticulous investigation and survey on the original architecture and the environment from different perspectives, in many aspects and at different time points. Furthermore, in-depth analysis and careful work was carried out by these designers in respect of the interior structure of the architecture. However, some suggestions shall be made as follows: First of all, more considerations should be given to consistency regarding the design of internal space of each garth since the four garths are connected, since they are design works in nearly the same period considering their architectural styles and of the similar design techniques. Although the functions of these typical architectures are different, the interior design style of them could be made under uniform arrangement. People may not feel that they're in an entire block of connected similar architectures as major differences are shown between interior styles of these architectures in terms of both building materials and design elements. In the second place, it is a pity that part of the contents (e.g. extraction of folk custom and elements of life) found out by the designers through early investigation and survey is not embodied in the later design.

To sum up, consideration put on the functional space of the base in the design work, the specific layouts as well as the clear transportation flow lines, all indicate the solid basic expertise of the designers as well as their capabilities and skills of problem analysis.

Gu Qiulin

专家点评

基地选择近现代哈尔滨快速城镇化过程中非常具有代表性的中华巴洛克历史建筑群，高度概念性地提炼出了"里里·里院"空间形态传统，通过详实的历史信息调研，以"大清官帖银行""黑龙江评剧院""松光电影院"历史建筑为核心，在充分尊重历史内涵的基础上，采用具有时代特征的建构语言，延续了"北三市场"的集体记忆。

朱海玄

The Chinese Baroque historic architectural complex, which features the fast urbanization in modern Harbin, is selected as the site, from which the traditional space form of "triple garth" is refined highly conceptually, and the collective memory of the "Beisan Market" is continued in an architectural language with the characteristics of the times with the historically built "Qing Dynasty Government Bank", "Heilongjiang Pingju Opera Theatre" and "Songguang Cinema" at the core based on sufficient historical information and full respect for the historical connotation.

Zhu Haixuan

青龙胡同酒店设计

CIID"室内设计 6+1"2017（第五届）校企联合毕业设计
CIID"Interior Design 6+1"2017(Fifth Session)University-Enterprise Cooperative Graduation Design

高　　校：	北京建筑大学
College：	Beijing University of Civil Engineering and Architecture
学　　生：	孙卫圣　尤昀　张乐情
Students：	Sun Weisheng You Yun Zhang Leqing
指导教师：	杨琳　朱宁克
Instructors：	Yang Lin Zhu Ningke
参赛成绩：	营造技艺与展陈组二等奖
Achievement：	Second Prize for Construction Technology Inheritance and Display Design

尤　昀
You Yun

张乐情
Zhang Leqing

孙卫圣
Sun Weisheng

学生感悟

　　大学的最后一次设计，我们竭尽全力用四年所积累的点滴证明自己。我们达成一致要在这次设计里为胡同注入活力，而不是一味的修缮和复原。胡同不是古董，时时刻刻影响着每一个居民的生活，它的居住功能从未消失。广州华南理工之行，不同地域的同学带着对自己地域老建筑的热爱分享着自己团队的调研成果和想法，每一栋老房子都值得被爱护和发掘。接下来的杭州之行也使我们受益匪浅。最后，我们在本校进行最后的答辩环节。精品民俗酒店的形象已经越发得清晰。做活人流导向，更多原本居民之外的人将会给胡同带来新的影响和活力。总之参加这次"6+1"校企联合毕业设计让我们收益良多。

Students' Thoughts

　　Because it is the last design of our university life, we make every effort to use four years ability to prove ourselves. We reach an agreement in this design to injeet vitality into Hutong, rather than blindly repair and recovery. Hutong is not antiques, it always affect the lives of every inhabitant, its living function has never disappeared. Guangzhou South China science and technology , students from different regions with their own old building's love to share their own team of research's results and ideas, each old house is worthy of love and excavation. The next trip 's, we also benefited. Finally in our school we have the final and defense. The face of fine folk hotel has become more and more clear. Designing the live flows, we have gain a lot will bring new impact and vitality to the alley. In short to participate in the 6 +1 school enterprises together to bring us a lot of income.

复原过程 ‖ Recovery Process

　　将现有青龙胡同街区道路及建筑现状进行图纸整合，分析区位因素得出适合的地块。经过分析，选择青龙胡同街道二级道路上一"L"形地块西侧的方形区域作为设计的整体选址。

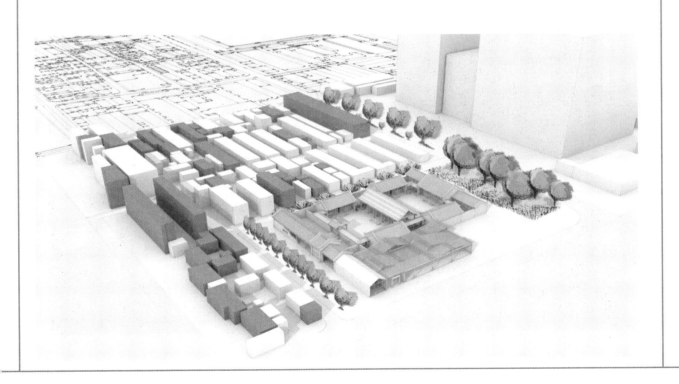

倒座改造 ‖ Inverted Seat Transformation

区域内部多为传统民居建筑，建筑层数多为一层或两层，自建房屋，安全性和耐久性均较差。胡同内传统民俗文化氛围浓郁，原有规划整齐，为南北向公共交通道路、南北向房屋以及单个独立的院落组合。

原有复原四合院为"井"字形，将"井"字形的中间部分院落保留，四周八个"方格"进行新的规划设计并将其改为室内空间。

51%的空间院落公共活动空间供顾客休息交流使用，主要陈设设施为座椅和茶几。27%的空间作为室内住宿空间如客房区，是整个民宿建筑空间最主要的功能区。22%的空间作为院落景观空间如主要种植植物的区域，作为景观观赏使用。

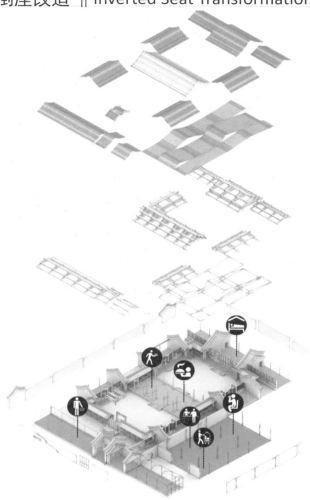

功能分区 ‖ Functional Division

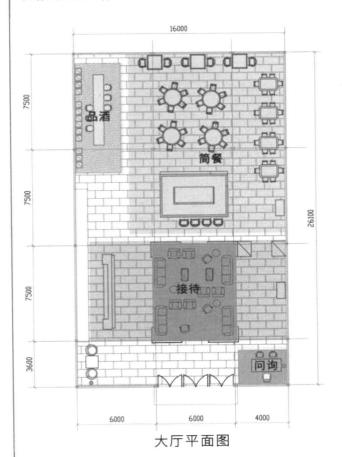

大厅平面图

整体的功能分区主要为由入口至庭院内部依次排列，其功能区越来越私密，封闭程度越来越高。所以，门厅和休息厅分布在入口处，伴随简餐区和员工更衣室、休息室等，用以和顾客流线尽量分开，保证建筑内部空间的完整性和私密性，同时给顾客更好的住房体验。客房区分布在院落的深处，为更进一步地利用和改造整个四合院建筑的格局，将原有院落改造成为由客房区包围的公共活动区。

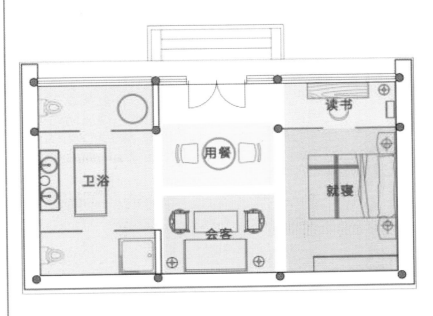

民宿建筑的公共性介于公共建筑和民居建筑之间，它既需要有公共建筑的开放性和包容性，也需要有民居建筑的相对私密性和归属感。总的来说，民宿建筑的空间形式主要需要以下几种空间功能来体现：

（1）门厅／休息厅。
（2）休息区。
（3）简餐区，主要功能是提供早餐和简便的下午茶等。
（4）客房区，占地面积最大，占有空间百分比最大，是整个民宿建筑空间最主要的功能区。
（5）后勤人员和服务区，供员工日常工作、休息使用。

答辩展示·民居建筑保护与更新设计

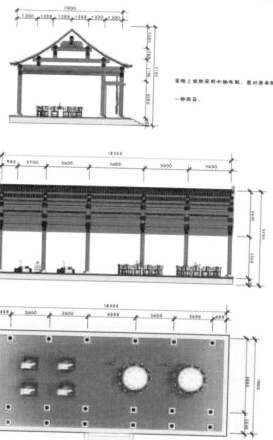

通过观察和图像了解青龙胡同周边情况。沿途进行雍和宫大街沿街商业形态和居民行为的大体观察，雍和宫大街沿街一大段主要为宗教周边产业和一些低端零售业，沿途有些居住环境较差的夹缝式居民空间，有一定改造价值，同时能看到不少大叔大妈沿街遛弯遛狗和一些聚集的大叔下象棋，由此可以想象，该区域缺少公共聚集的娱乐场所供居民使用，因为大街并不适合公共娱乐。之后由北新三条进入北新胡同和炮局胡同等连片的胡同区域，内部交通、商业较为杂乱，但比较稳定，室内居住环境和基础设施有需要完善的地方。

答辯展示·民居建築保護與更新設計

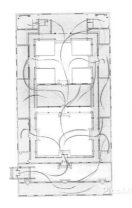 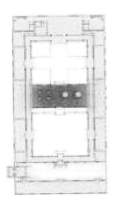

雍和宫大街西侧有五道营和方家胡同，两条胡同脉络很清晰，商业化做得很好，以五道营胡同为最。综合考虑两个方向，公共娱乐空间和局部居民居住空间，另外还要考虑到当地人固有的生活需求。

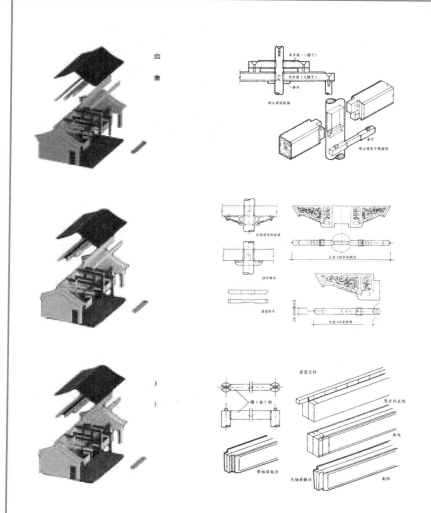
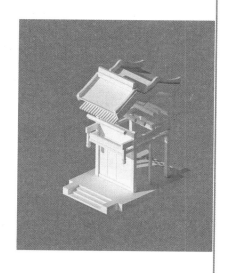

导师点评

该项目能从传统胡同内建筑形制及传统木构建筑结构关系入手,在酒店室内布局、展陈设计中能充分展现传统元素与营造法式的继承与沿用。整体改造方案图面表现完整,设计图解系统有一定逻辑。但在受众分析、对调研成果中发现问题及如何去解决问题上欠缺设计表达。

任彝

The project proceeds with the relationship between the architectural form of the traditional alley and traditional wood architectural structure, brings into full play the inheritance and constant application of traditional elements and methods of construction in respect of interior layout and displaying design in the hotel. The drawing of overall reconstruction scheme is complete in presentation, and certain sense of logics is reflected in the graphics system of the design. However, the project is weak in design expression, including audience analysis, problem finding in the investigated results and solution to such problems.

Ren Yi

专家点评

传统民居改造成精品民宿酒店是对传统民居的有效利用,不论是空间体验还是空间的活化,都让人体验到传统民居的文化内涵。设计者对院落环境和室内空间进行酒店化改造,设计了不同的酒店房型,调整了空间序列,在室内空间增强了文化性、艺术性的表达,家具、陈设和空间的色彩、材料都经过了认真的思考和挑选,营造出雅致的居住氛围。与此同时,设计者采用尊重传统建筑建构的思路,细致分析了木结构屋架的营造作法,使得新建建筑与老建筑和谐共生。不足之处是设计者对酒店的体验较少,对精品酒店的经营方式理解不够,因此空间的组织和功能布局存在缺陷,例如设计中缺少庭院景观的内容,建筑与气候的适应性不足,客房使用功能不够完善。

张磊

Reconstruction of traditional dwellings into a boutique home-stay hotel is the effective utilization of these dwellings, where tourists are able to experience the cultural connotation of these traditional dwellings. The designer makes reconstruction for courtyard environment and interior space for hotel purpose, designs different types of hotel rooms, changes spatial sequence, and promotes cultural and artistic expression in the interior space. Serious consideration and careful selection are given to furniture, furnishings, colors of space and materials for creating an elegant living atmosphere. At the same time, the designer respects the traditional architecture construction, and meticulously analyzes the construction method on timberwork roof truss, making harmonious coexistence of newly-constructed buildings and old ones. The deficiency of the design lies in spatial organization and functional layout because the designer has limited experience of hotels and is not familiar with operation of the boutique hotel. For example, in this design work, content of courtyard landscape is not adequate; the architecture fails to meet the requirement concerning the changing climate; functions of the guest rooms are incomplete.

Zhang Lei

拾遗——南京市非物质文化遗产展

CIID"室内设计 6+1" 2017（第五届）校企联合毕业设计
CIID "Interior Design 6+1" 2017 (Fifth Session) University-Enterprise Cooperative Graduation Design

高　　校：	南京艺术学院
College：	Nanjing University of the Arts
学　　生：	林秋霞　王依丽　章伶钰
Students：	Lin Qiuxia　Wang Yili　Zhang Lingyu
指导教师：	朱飞
Instructors：	Zhu Fei
参赛成绩：	营造技艺传承与展陈设计组二等奖
Achievement：	Second Prize for Construction Technology Inheritance and Display Design

林秋霞
Lin Qiuxia

王依丽
Wang Yili

章伶钰
Zhang Lingyu

学生感悟

作为展示设计应届生，我们很荣幸参加本届的"6+1"校企联合毕业设计，从三月开题到六月答辩，从广州到杭州再到北京，我们跟其他几所学校的同学以及老师们共同走过了一段难忘的旅程。每次的相聚，都使我们收获颇多。本次关于传统民居的保护性利用设计的这一命题，我们最终将南京甘熙宅第作为我们设计的对象，期间也研究了许多相似案例，感受到了传统民居在设计师的手中绽放出了新的生命力。此次毕业设计虽有遗憾但更多的是成长，是结束同样也是开始。

Students' Thoughts

As the of display design, we are honored to be invited to take part in this "6+1" University-Enterprise Cooperative Graduation Design. From the establishment of our graduation project in March to the completion of our thesis defense in June, we travelled from Guangzhou to Hangzhou, and then to Beijing. We have worked with teachers and students from several other universities, shared unforgettable memories with each other, and gained a lot in each meeting with them. We ultimately took Ganxi Mansion in Nanjing as our project subject based on the project theme concerning the protective utilization of the traditional dwellings. During the implementation of the project, we made research on a large number of similar cases, and could feel the new vitality brought out by the design of the traditional dwellings. Although our graduation project is not perfect, we learned a lot. The completion of the project also signifies that we still have a long way to go.

展区规划 ‖ Exhibition Planning

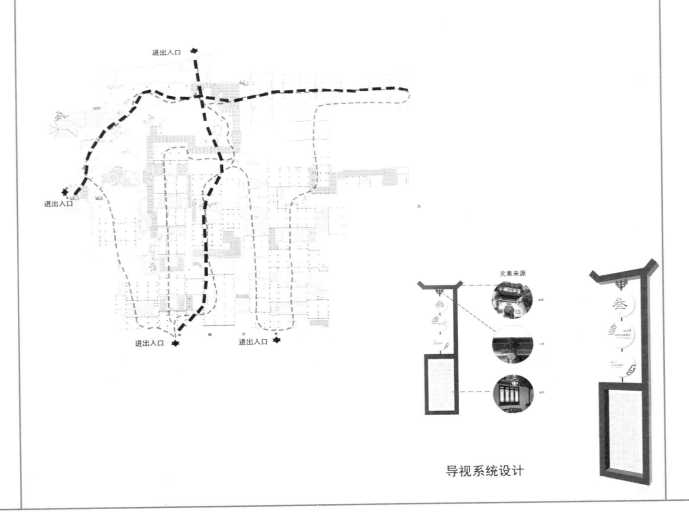

导视系统设计

新建筑展厅设计 ‖ New Exhibition Hall Design

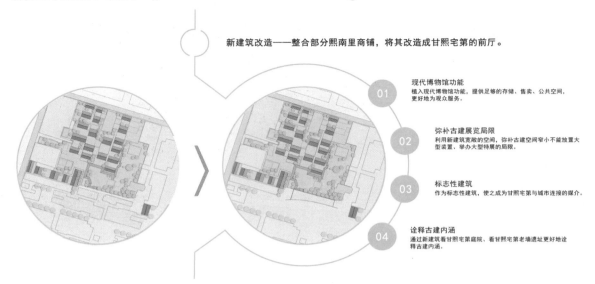

新建筑改造——整合部分熙南里商铺，将其改造成甘熙宅第的前厅。

01 现代博物馆功能
植入现代博物馆功能，提供足够的存储、售卖、公共空间，更好地为观众服务。

02 弥补古建展览局限
利用新建筑宽敞的空间，弥补古建空间窄小不能放置大型装置、举办大型特展的局限。

03 标志性建筑
作为标志性建筑，使之成为甘熙宅第与城市连接的媒介。

04 诠释古建内涵
通过新建筑看甘熙宅第庭院、看甘熙宅第老墙遗址更好地诠释古建内涵。

地下一层展厅设计 ‖ Underground Exihibition Hall Design

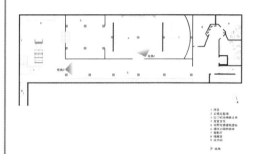

皇室文化：
以南京云锦长期用于专织皇室龙袍冕服，是皇室贵族的象征，此空间以皇家代表色黄色作为展厅主色调，以龙纹柱为视觉中心，结合四周展墙上的图版、展柜里的云锦实物进行展示，地面与墙面隐约显现的云纹既呼应主题也进一步彰显皇家气质。

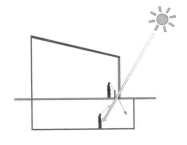

阳光透过一楼狭长的镂空通道照射到负一楼，同时在负一楼甘熙宅第建筑遗址旁的过道里隐约可以看到甘宅小园里的景色。

甘熙宅第的建筑遗址（视角2）

一层展厅设计 ‖ First Floor Exhibition Hall Design

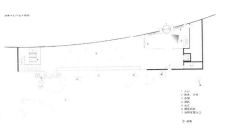

新旧建筑连接处，保留了甘熙宅第原入口造型，使其成为新馆的一部分，同时也保护了入口环境。往右的空间放置了一些自由排列的展台，新馆与甘院庭院相接的地方采用了透明的落地玻璃墙，使得新旧建筑联系更加密切，观众在新馆即可欣赏到庭院全景。

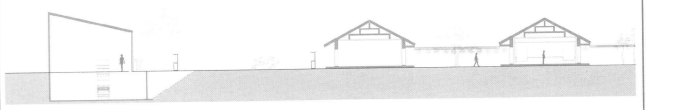

秦淮灯会展厅设计 ‖ Design of Qinhuai Lantern Hall

结合古建中轴对称并充分利用原建筑进深较深、层高较高的特征，以层层递进的空间关系与建筑本身相融合。展示柜外利用挂落等软装营造空间效果，柜内集中展示各类花灯。

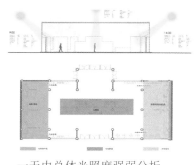

一天内总体光照度强弱分析

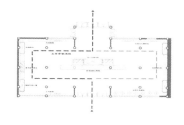

秦淮灯会展馆人流动线图

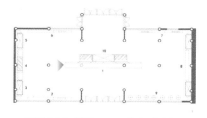

1 秦淮灯主题墙　　6 花灯装置
2 花灯装置　　　　7 花灯文化与民俗
3 传统花灯手艺　　8 多媒体互动
4 秦淮灯会历史　　9 各类花灯实物
5 花灯类别　　　 10 名人花灯作品

经书佛缘展厅设计 || Design of Buddha Book Exhibition Hall

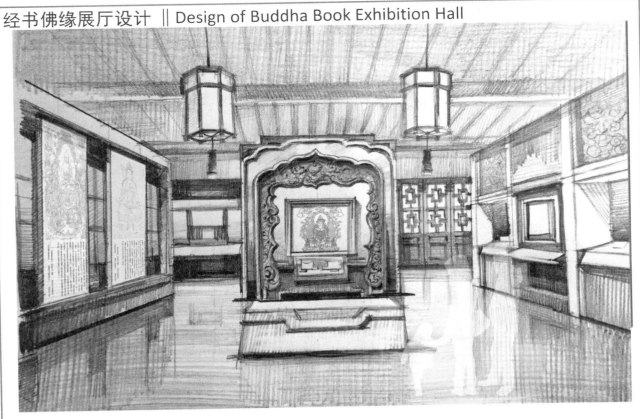

抓住"佛"与"刻"这两个关键字，在展厅与展柜的设计上融入佛教元素与雕刻的元素，使传统的古建空间与现代化的展示手法有机结合，营造古意盎然，传统色彩浓郁，却不失现代感的展厅环境。

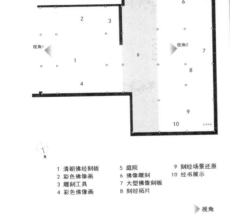

1 清朝佛经刻板　5 庭院　　　　9 刻经场景还原
2 彩色佛像画　　6 佛像雕刻　　10 经书展示
3 雕刻工具　　　7 大型佛像刻板
4 彩色佛像画　　8 刻经拓片

▶ 视角

厅内厅外充分利用原有的园林空间，采用借景、对景的手法，营造清幽素雅的休憩环境，使厅内厅外有机融合。

一体化展柜设计 ‖ Showcase Design

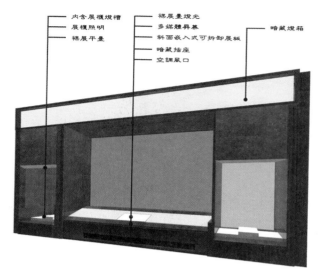

通过对可移动展柜的再设计将大量功能整合到了展柜中，有效的消隐了部分基础设施带来的管线外露等问题。室内照明、展板灯槽、裸展台灯光、踢脚灯、空调风口、音响设备、多媒体屏幕、插座等常用设备器材皆置于展柜中。

一体化展柜高度集成化，建筑内不开槽，不吊顶，不隔墙，不破坏，将展柜系统，照明系统等融入展柜内，满足当今各种展陈新需求，将古建的保护臻于极致。

沿建筑轮廓，与展柜系统紧密结合，形成一个完整的基础设施系统。由中央管理机器统一控制。主展厅、临展厅等空间相互串联，便于管理和维护。

导师点评

传统民居用作非物质文化遗产的展示空间是很有意义的设计课题。该作品结合建筑空间设计、自然光照分析、导视设计、家具设计进行室内展示设计。效果图表现技法娴熟，特别是手绘表现力强。观众能从展示中感受到文物的美感，激起更多兴趣了解南京文化。

建筑功能分区明确，但是从游客中心到国家非遗展示区的路线不明确，两者之间应有直接便捷的联系。游客中心负一层甘熙宅第建筑遗址的保留充分显示对古建筑的尊重，一楼的观赏空间若扩大一些，可更好地与负一层形成共享空间。秦淮灯会展厅的效果图隐藏了结构柱子及其他构造物，未能真实反映空间关系及布局的合理性。"经书佛缘"手绘效果图表达的空间与平面图无对应关系。

薛颖

It is a meaningful design topic to use the traditional dwelling as the display space of the intangible cultural heritage. This work presents the interior display design through the integration of architectural spatial design, natural lighting analysis, guided design, and furniture design. The presentation technique of the renderings, especially for freehand drawing, is professional. From the exhibition of this design, the audiences are able to experience the aesthetics of the cultural relics, thus arousing more enthusiasm of learning culture of Nanjing.

While the building functions are defined clearly, the route linking tourist center and national intangible cultural heritage display area is clearly specified, therefore a direct and convenient route needs to be established. Preservation of building ruins in the Ganxi Mansion on the basement floor at the tourist center shows much respect to the ancient architecture. If the sightseeing space on the first floor is larger, the shared space may be created with that of the basement floor. Structure pillars and other structures cannot be seen in the rendering exhibited in the lantern exhibition hall in Qinghuai region. Consequently, the rationality of the spatial relationship and layout cannot be presented truly. The space presented in the freehand sketching Confucian Classics and Buddhism Connection and the plan have no corresponding relation.

Xue Ying

专家点评

该组同学的设计突出新旧融合和展柜一体化，较好地把握了文保建筑、传承和现代展示设计表现之间的关系，以及传统民居与新建筑在功能、内容和形式方面统一对比。在空间组织、内容信息、造型形态、技术设施设备兼容性方面具有一定探索。对提高广大群众学习非遗、保护非遗的热情，同时也对南京市非物质文化遗产的研究、展示、保护以及引导观众思考南京非遗如何在当代全球化环境下进行继承与创新具有一定的借鉴意义。

曾军

The work designed by the students in Team One highlights old-and-new fusion and showcase integration, grasps the relationship between cultural protective architecture, inheritance and modern display design presentation in a good manner. Uniform comparison is made between the traditional dwellings and the new buildings in the aspects of function, content and form. Spatial organization, content information, modeling form, technical facilities and equipment compatibility are, to some extent, reflected in the design work. It is conducive to the enthusiasm of the broad masses to learn the knowledge of intangible cultural heritage and protect it, and has certain reference significance for research, display and protection of intangible cultural heritage in Nanjing, as well as for guiding the audience to think about how to conduct inheritance and innovation in the context of the contemporary globalization.

Zeng Jun

围炉共笙
——浙江省台州市黄岩半山村传统营造技艺传承及展示设计

CIID "室内设计 6+1" 2017（第五届）校企联合毕业设计
CIID"Interior Design 6+1"2017(Fifth Session)University-Enterprise Cooperative Graduation Design

高　　校：	浙江工业大学
College:	**Zhejiang University of Technology**
学　　生：	陈玫宏　胡安琪　洪朝艳　杨洁
Students:	Cheng Meihong　Hu Anqi　Hong Chaoyan　Yang Jie
指导教师：	任　彝
Instructors:	Ren Yi
参赛成绩：	营造技艺传承与展陈设计组三等奖
Achievement:	Second Prize for Construction Technology Inheritance and Display Design

陈玫宏
Chen Meihong

胡安琪
Hu Anqi

洪朝艳
Hong Chaoyan

杨　洁
Yang Jie

学生感悟

很荣幸参加这次的CIID "室内设计6+1" 2017（第五届）校企联合毕业设计活动，并且与其他学校学生切磋学习，通过系统的前期调研到分析设计地块的现状并提出问题，再到最后提出概念，并且设计完成整个项目，我们提升了各个方面的能力。加之，古村落的改造与保护也是如今的热点问题，通过对于半山村的规划、改造方面的设计，我们更加系统地接触了这个领域，认识到设计是以人为出发点，将建筑形式依附于建筑内容。这次的毕业设计让我们感触良多，受益匪浅。

Students' Thoughts

We are honored to be invited to take part in CIID "Interior Design 6+1" 2017 (the 5th session) University-Enterprise Cooperative Graduation Design and learn from peers from other universities by exchanging views. We raised questions through early systematic investigation and analysis on current status of the designed plot, then came up with the concept and finally designed and completed a project, through which our abilities were improved in various aspects. Furthermore, the reconstruction and protection of ancient villages has been a hot issue attracting much attention today. Through planning and reconstruction design for Banshan Village, we were involved in this field in a more systematic way. We realized that design shall be based on people at the very beginning of the design process; the architectural style depends on the architectural content. We had mixed feelings and gained a lot from the implementation of the graduation project.

答辯展示·營造技藝傳承與展陳設計

深入調研 ‖ In-depth Research

半山溪視線分析

地理環境 建築形式

地方材料 空間形式

文化底蘊 空間內容

我們分別從切實了解現狀、了解村民訴求及感知地方語言對台州黃岩富山鄉半山村進行了詳盡的調研，旨在全面理解半山村優劣及其特色，設計出有益半山村發展，且具半山村生命力的作品。

C建築西南破敗
C建築西南破敗
C建築西南破敗
C建築西南破敗
C建築西南破敗
C建築西南破敗
C建築西南破敗
C建築西南破敗

設計地塊位於半山村中上位，臨近半山溪，視野良好，是由七棟民用建築組成的建築群，建築破敗不已，建築風格也隨建造年代不同而風格不一。

概念形成 ‖ Concept Formation

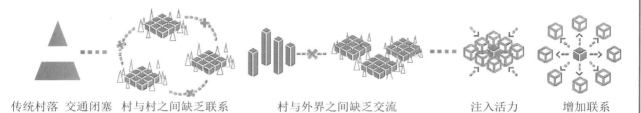

综上所述，考虑到半山村的种种劣势，我们想营造一个围合、温暖、共同，热闹的场所，人们围着炉火在一起，音乐歌舞、热闹非凡，寄托对半山村民美好生活的祝愿：鼓瑟鼓琴，笙磬同音。

具体方案 ‖ Specific Plan

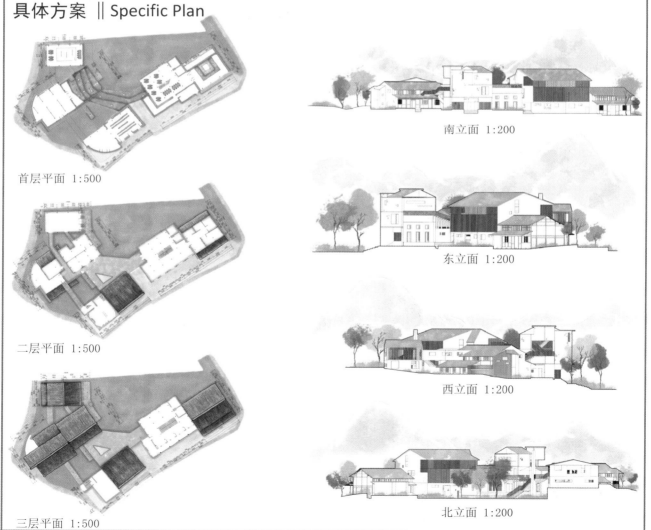

建筑更新 ‖ Building Updates

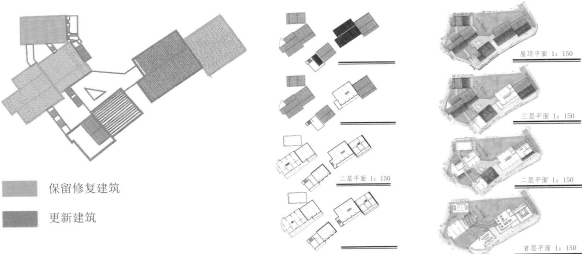

保留修复建筑

更新建筑

一层空间 ‖ First Floor Space

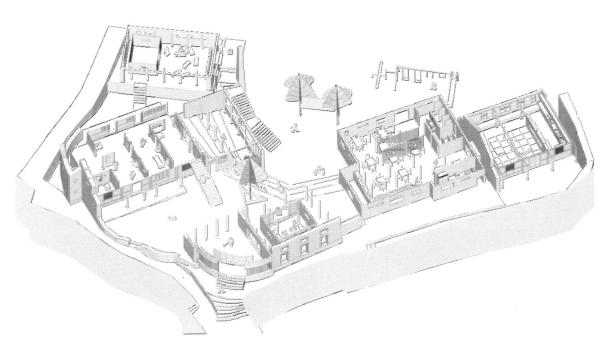

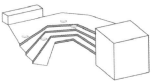

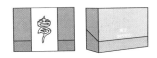

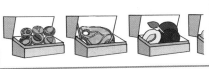

一层空间为活态传承体验馆，旨在区别于以现代科技手段对非物质文化遗产进行"博物馆"式的保护，用文字、音频、视频等方式记录非物质文化遗产项目的方方面面。

答辩展示·营造技艺传承与展陈设计

二层空间 ‖ Second Floor Space

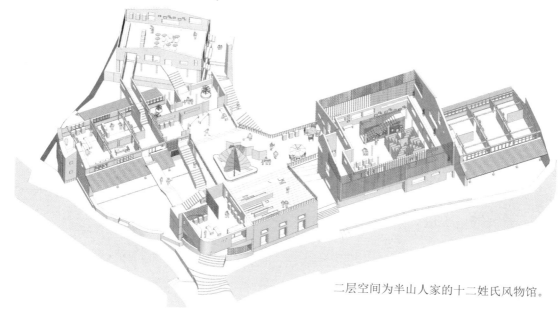

二层空间为半山人家的十二姓氏风物馆。

三层空间 || Third Floor Space

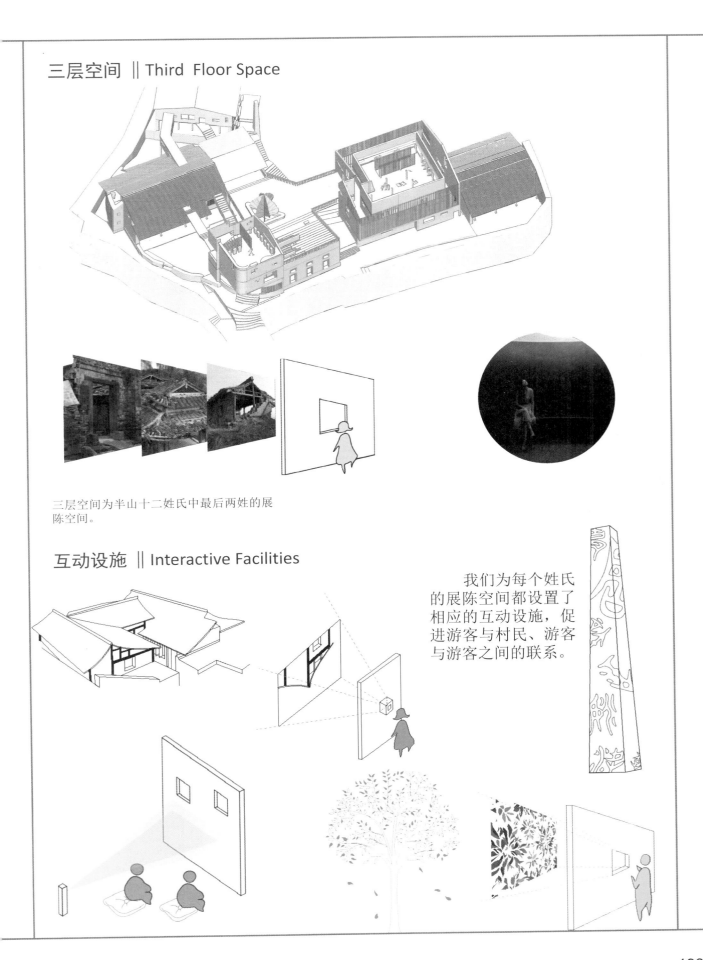

三层空间为半山十二姓氏中最后两姓的展陈空间。

互动设施 || Interactive Facilities

我们为每个姓氏的展陈空间都设置了相应的互动设施，促进游客与村民、游客与游客之间的联系。

环境标识 ‖ Environmental Sign

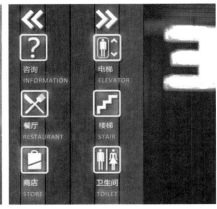

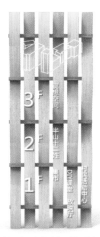

结合半山村的各种元素，设计了相应的环境标识系统。设计竹管灯，照亮半山村的夜晚。

由于半山村在地理位置上处于半山腰，交通并不便利。为此我们针对这个问题设计一个专属半山，方便村民与游客的应用APP。同时在APP上我们还设置了手工艺课程的预约教学，促进外来游客对半山村的传统技艺的了解，同时使传统技艺得到有效的传承。

导师点评

通过对半山村传统村落的设计改造，将传统元素有机地植入到改造空间内，同学对空间的理解较为深刻，能将传统院落空间转变成符合现代使用功能的展览空间。同时改造后的空间环境中又具有一定的传统文化属性。

空间的传统文化继承、保护与利用，应建立在文化层面上，应是一种文化的"软"继承，不应该一味追求传统形式上的复制与罗列，应做到精神层面上的继承与发扬，这样的保护与利用才能良性发展，故此，同学们在传统村落的保护与利用研究方面要站在更高层面上去思考，这样对未来、现在、过去都是有益处的。

马辉

Traditional elements are embedded into the space through design and reconstruction of the traditional settlements in Banshan Village. The students have such a good understanding about the space that they are able to change the traditional courtyard space into an exhibition space meeting the requirement for modern application. The space environment also has certain traditional cultural characteristics after reconstruction.

Inheritance of traditional culture, protection and utilization of the space shall be subject to culture aspect. Such cultural inheritance shall be in an "organic" manner, where the protection and utilization may be put into operation in a good way through spiritual inheritance and promotion other than traditionally blind pursuit of reproduction and listing. Consequently, the students shall ponder over the research on the protection and utilization of the traditional villages from a deeper perspective, which is good for the past, present and future.

Ma Hui

专家点评

在浙江美丽乡村建设中，如何突破乡村看景尝鲜的习惯模式？如何在一个过路的小村留住客群？浙工大毕业生创作的《围炉共笙》给我们带来了一份独特的答卷。

通过集会、方言、姓氏、科普等内容，设计小组抓住围炉的主线，层层递进、环环相扣，力图营造一个传统与时尚共生的场所。我认为，这种跨界思考的乡旅项目创意及同学在衍生产品上的发力，都使作品达到了优良、原创的水准。

王炜民

How to change the old habit of trying new things concerning sightseeing in rural area during the construction process of the beautiful village in Zhejiang? How to retain customers in the village where they travel? These questions are answered in a unique way by the design work Conversation Surrounding the Furnace by the graduates from Zhejiang University of Technology.

Through rally, dialect, surname, dissemination of science and other content, the design team grasps the main line of the design theme surrounding the furnace, and makes progress step by step and interlocking of each link so as to strive to create a place where the tradition element is able to exist together with the fashion element. The design work, I believe, has reached an excellent and original standard on account of the originality of the rural tourism project and the power generated by the students for the derivative products.

Wang Weimin

匠心椅阑杆

CIID "室内设计 6+1" 2017（第五届）校企联合毕业设计
CIID "Interior Design 6+1" 2017(Fifth Session)University-Enterprise Cooperative Graduation Design

高　　校：	南京艺术学院
College:	Nanjing University of the Arts
学　　生：	赵雨琦　朱彦　黄雅君
Students:	Zhao Yuqi　Zhu Yan　Huang Yajun
指导教师：	朱飞
Instructors:	Zhu Fei
参赛成绩：	营造技艺传承与展陈设计组三等奖
Achievement:	Third Prize for Construction Technology Inheritance and Display Design

赵雨琦
Zhao Yuqi

朱彦
Zhu Yan

黄雅君
Huang Yajun

学生感悟

　　此次参加"6+1"比赛的经历是令我们难忘的，短短几个月，奔波了很多，学习了很多，更认识到了我们的诸多不足，对于设计有了更加深刻、透彻的理解。而这一次毕业设计的主题和我们展览的选题是我们从未涉足的崭新领域，古老的传统、悠久的文化、精湛的工艺和匠人的情怀都让我们有所感慨并且收获良多。

Students' Thoughts

　　It was an unforgettable memory of us to participate in this 6+1 competition. Over the past few months, we travelled to many places and learned a lot. More importantly, we have realized more about our weakness, and had a deeper and more thorough understanding of design. The theme of the graduation project and the subject we selected for our display design is completely new to us. We have learned a lot from the ancient tradition, long-standing culture, fine workmanship as well as the passion of the craftsmen.

古建筑规划 || Ancient Architecture Planning

总平面图
- 积厚流光
- 巧夺天工
- 匠心独运
- 大工精诚
- 安居而坐
- 新馆建筑
- 临展区&工作区

节点规划图　　路线规划图

答辩展示·营造技艺传承与展陈设计

北欧现代坐具展 ‖ Scandinavian Modern Chair Exhibition

在负一楼的丹麦坐具展上，我们把丹麦名设计师的坐具作为展品，在折线的展台及展板上通过图文描述、多媒体播放以及还原丹麦代表性窗台的烘托氛围的方式展出。展览和甘家大院内部的明清坐具展览相呼应、相对比，体现传承与创新。

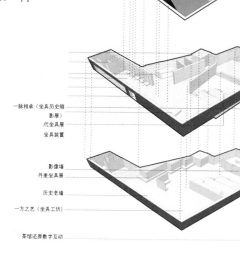

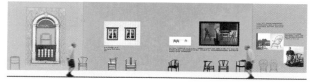

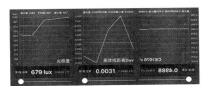

展区布局 ‖ Exhibition Layout

匠心独运

因为甘熙故居本身是一个阶级制度严谨的宅院，我们大体沿用了中轴对称的方式分隔展览区域。在中轴上部、下部分别是两个互动区域，在展厅左右两侧分别是异形展台和以屏风为背景的微缩模型展示。

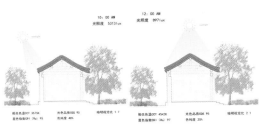 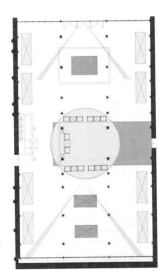

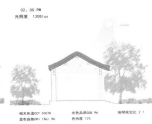

光照分析图

榫卯是中国木建的重要基本制作技术之一，明清家具几乎运用了所有榫卯结构种类中的精华形制。

在临展厅中有着这样一个巨大的榫卯装置，是由椅子腿的部件解构而来，下面环绕着展台，同时垂挂着各种精巧的榫卯构件供参观者观察欣赏。

榫卯结构深刻的文化内涵不只表现在木构家具的拼接中，更体现了古代中国科技与人文的结合。

这一部分的展台是不规则切割的，展台上放置微型椅子模型，展板兼具图像放映和图版介绍的功能。

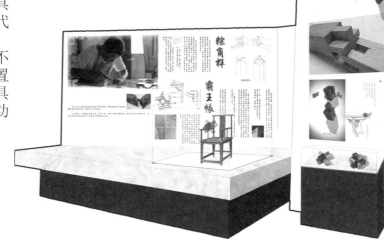

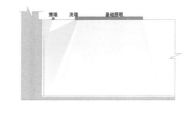

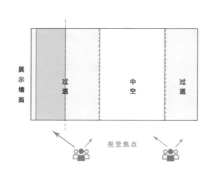

展馆建筑形式 ‖ Architectural Form of the Pavilion

大工精诚

甘熙故居原非物质文化遗产三展馆的展示区域是通过解构具有民国风情的西洋建筑,再加以重组,窗台、房檐、墙沿,建筑的各种构件都可以成为椅子展示的平台,这样既是对一定历史时期南方风貌还原,又将椅子构筑在不同的时空中,使得椅子更加具有历史的厚重与积淀。同时非物质文化遗产三展馆中间有一处暗房里面变幻着各种各样的光影来烘托椅子。

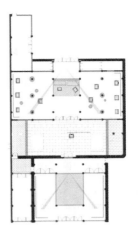

尺有节目

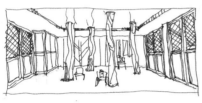

原先的非物质文化遗产四展馆没有充分利用原先的建筑特色，馆内展陈方式陈旧，展品积灰，光线昏暗，展示效果不佳。

在甘熙故居原非物质文化遗产馆四展馆的部分，展示制作明清家具木材用料的选定和讲究。在这个受到诸多现实条件制约、但是面积也是最大的房间里，利用原先的地形、布置、建筑和光照特色，将梁柱变为森林，将展台变为树桩，将窗花变为树叶，阳光透过窗户洒进来仿佛树影婆娑，椅子在树丛中仿佛天然生长。这样的设计主要是为了表现自然，表现制作椅子的原材料在自然中的生态。每把椅子都由它身边的这种类树木制作而成。

注重互动性和体验性，椅子摆在树木边，旁边没有任何展牌标识，而玄机正在树木里，通过绑定在树木上的微型投影仪向树洞投影信息。树洞代表着秘密，关于制造椅子的原古树的秘密由着参观者自己去窥探。另一样重要的互动体验形式是自然之书。书页也是从树木而来。参观者进入展厅前可以领取书板一块，当进入展厅，走到右手边的墙沿，站在特定记号标注的位置时，感应系统捕捉到书板的存在并随机打开位于天花的投影，向书板上投下相应的关于制作椅子所需要的木材的相关知识。

导师点评

　　展示设计融合了空间元素、视觉元素、材料应用及照明设计等领域的设计语汇,已经不仅仅是对产品进行展览形式的表达。现代展示设计要求设计师在设计构思环节,必须结合展示对象在"如何陈列?"环节上作出设计师个性化的表达,才能使设计主题、设计思想的创新更趋于完善。

　　南京艺术学院朱飞老师指导的"匠心椅阑干"展示设计,把古典元素进行分解、提炼、重构,使展示设计呈现出电影镜头的蒙太奇趣味,把解构主义的视觉元素呈现方式与空间相结合,在展示"方式"、展品的陈列与空间气氛营造的研究环节作出了创新,并把中国元素拓展至视觉导示系统的应用。

　　相对于设计整体的立意,图面表达方面稍显逊色,尤其对于展厅空间设计表现方面略简单,没有把"匠心"转换为视觉元素展现在"空间"中,展品与空间的关系方面细节的表达不足(仅停留在形象符号的获取与利用),设计细节对于主题的呼应环节留有遗憾。

宋扬

Spatial elements, visual elements, material application, lighting design and other industrial design vocabulary are incorporated into the display design, giving an expression not simply for product display. The innovation in theme and idea of the design will not be more perfect until individual expression are made available for object to be displayed in the "how to display" link by the designer during the design concept process according to the requirements of modern display design.

The display design "complexity-based ingenuity", mentored by Zhu Fei, a teacher from Nanjing University of the Arts, conducts decomposition, refining and restructuring of classical elements, indicating a montage fun of display design in the film. In this design, the integration of presentation method and space of visual elements of deconstructivism brings innovation in respect of displaying "method", exhibits display and creation of spatial atmosphere, expanding the application of Chinese elements to visual guide system.

Compared to overall conception of the design, presentation in drawing is somewhat inferior. In particular, the presentation for spatial design of the showroom is so simple that it fails to convert "ingenuity" into visual elements in the "space". Shortcomings are also shown by the details presentation in relating to the relationship between the exhibits and space (only in the obtaining and usage of the pictorial symbol). It is a pity that the design details fail to bring out the crucial points of the design theme in the corresponding links.

Song Yang

专家点评

　　该组同学以屏风式的写意手法表现中国文化,在展陈道具的设计上采用构建式的手法结合传统元素,新式新颖,仪式感较强,一些想法和形式给甘家大院的保护性利用带来很多启发。传统民居的保护性利用需要文化的沉淀,一切形式要依附展览的内容,而同学们在解决实际问题方面还有所欠缺。

曾军

The students in this team present the Chinese culture with screen-based freehand method. As for design in the displaying props, constructive method is adopted with integration of traditional element, which is novel in style and strong in sense of ceremony. Some ideas and forms have provided much inspiration for the protective utilization of Gan Family's Courtyard. The protective utilization of traditional dwellings requires the accumulation of culture. All forms shall be based on the displaying content. However, it is shown that the students are weak in solving practical problems.

Zeng Jun

教学研究
Teaching Research

CIID『室内设计 6+1』2017（第五届）
校企联合毕业设计
CIID"Interior Design 6+1"2017(Fifth Session)
University-Enterprise Cooperative Graduation Design

国匠承启卷
——传统民居保护性利用设计
Craftsmanship Heritage
——Design for the Protective Utilization of Traditional Folk Houses

國匠承啓卷——傳統民居保護性利用設計

Practice and Reflection on the Supervision to Graduation Design in 2017

Zuo Yan

The graduation design project of this year was carried out around the theme of dwelling conservation. The cooperative enterprise set the topic and provided guidance. Therefore, there is a big difference in guidance compared with before, which mainly lies in the following two points: the first is the nature of the topic, which involves the functional renovation of the historic architecture in Shanghai, as well as the interior and exterior renovation of the rooms concerned. This means strong comprehensiveness; the second is the students' ability and character, in which there is a great difference, which adds to the difficulty of the teamwork. This time, Tongji partnered with G-Art Design International, a large-scale local interior design company that rich industry experience, on the graduation design. The design director Mr. Huang Quan participated in setting the topic and offering guidance at the early stage. The project task had a high requirement for the teachers and students' practical operative ability.

1 About the Topic

The topic is focused on city micro-renewal, a hot issue that is drawing constant attention from the present society. The renovation object is the ground floor of the Blackstone Apartment building, which was built in 1924 and is located in Fuxing Middle Road, where the French Concession was once situated. The renovation content includes redevelopment of the overall building, involving functional planning, conservation and utilization of the architectural heritage, reconstruction of the building structure and exterior facade, landscape design, interior design of important functional spaces. The work involves a lot of interdisciplinary knowledge and skills, such as urban design, historic protection design, building renovation and landscape design, which means an extremely high requirement for the graduates' comprehensive ability and quality. The topic is a real city micro-renewal project, which has a concrete, practical requirement for design, and is a great challenge to the students' application of their knowledge system.

It is not easy to fulfill the design task for the following several reasons: first, the study of the history of the site is where the design starts. The Blackstone Apartment inside the site, which was built in the 1920s, is an excellent historical building in Shanghai, and as a modern premium apartment in Shanghai, it has gone through over 90 years of hardships, which can be seen from the change of inhabitants. Most of the first inhabitants were senior staff of foreign companies or highly-educated doctors and professors. After liberation and during the Cultural Revolution, there lived workers, government agents and red guards, with capitalists and senior intellectuals driven out. Today, there live people of different backgrounds, including a small number of senile permanent residents and a great number of tenants, such as foreigners and domestic migrants. The change trend of the residents' background is a basis for historical research and future function positioning. Apart from a thorough, comprehensive research of the site, including the present structure, orientation and usage of various buildings, a visit needs to be paid to the residents in and around the Blackstone Apartment to learn about their living condition and the former residents' lifestyle. Particularly, research needs to be carried out carefully since only the ground floor will be renovated, while the upstairs residents cannot be disturbed. The second is the positioning of the commercial function of the site. Functional analysis and business planning are of vital importance to a real project, for which there should be, generally speaking, a specific design specification. This time, the function positioning needs to be carried out by the students. In other words, the students need to make a design specification on their own. As we have planned, the site's function positioning must be carried out based on a sufficient, extensive research of the peripheral functions. Therefore, we clearly intended to carry out a research on function distribution in two regions with the radius of 1km and 500m respectively with the site at the center at the early stage of design. Also, we organized a visit to some commercial projects about historical architecture development such as Hengshan Yaji and Wukangting. Analyzing and cleansing the research data, we gave advice on how to enhance the cultural and artistic functions of the site; third, in the site there are 3 buildings, among which No.2 and No.3 still have a sturdy and durable structure, which just needs to be renovated in local parts such as the exterior facade. The south-facing open space in the Blackstone Apartment was a gorgeous garden, which has now been corroded by garages and talent apartments. About this, we need to make an in-depth discussion on how to put forward a renovation scheme for the building structure and a reuse scheme for the historic garden view using our architectural knowledge.

2 About Theme Interpretation

Considering the features of the topic, theme interpretation appears very important. The heuristic discussion carried out at the early stage of design helped the students clearly understand the design tenet, expanded the research vision, and enabled the students to use an effective research method to start design with the urban function and historic preservation in the first place. There was a difference between the two groups in research result and cognition, and they set a design objective and identified priorities respectively. Group 1 paid attention to the "lateral landscape", a common but noticeable feature revealed based on identification of many practical problems during research. From the intersection of the site entrance and Fuxing Middle Road to the site, other buildings in the site have an unnatural lateral relation with the main building—Blackstone Apartment, and this relation is left over from history. Whether it's valuable or not to form a design concept, it depends on how to alleviate this distorted positional relation. But this involves many conflicts caused by lots of historical reasons that have something to do with city renewal. In a sense, this relation cannot be rationalized by design alone, while the developer, designer, government and community need to work together to solve it. So, I hoped to present this practical site-related problem before the students to provoke their thinking about it. After much discussion, the groups began to make a design with art, experiment

2017年毕业设计指导心得与思考

左琰

今年毕业设计按学会的要求，课题需围绕民居保护话题，由相关企业出题和指导，因此指导起来与往届相比有较大不同，主要在两点：一是与课题性质有关，涉及上海历史地段保护建筑的功能改造及其场地里的建筑室内外改造，综合性高；二是学生，这次的学生在能力、性格等方面有较大的个体差异，合作成组需要磨合和适应。这次同济的毕业设计是与行业经验丰富、规模较大的一家本地室内设计公司——集艾室内设计公司合作，设计总监黄全先生作为毕业设计企业指导教师参与了命题和前期指导工作。此次毕业设计实操性强，为师生提出了更高的要求。

1 关于课题

这次选题聚焦当前社会持续关注的热点话题——城市微更新，以昔日法租界复兴中路历史街区中建于1924年的黑石公寓大楼的底层功能改造为契机，将场地内其他既有建筑作为一个整体一起进行功能再开发，完成功能业态策划、建筑遗产保护利用、既有建筑结构和外立面改造、场地景观设计、重要功能空间室内设计等工作，涉及了城市设计、历史保护设计、建筑改造、室内设计和景观设计等诸多跨学科领域的知识和技能，对毕业生的综合能力和素养要求相当高。这个选题作为真实的城市微更新项目对学生提出了具体实际的设计要求，对学生以往所学的知识体系是一个全方位的挑战和检验。

这次选题有以下几个难点。第一，基地的历史研究是开展设计的起点，基地内建于20世纪20年代的黑石公寓是上海市优秀历史保护建筑，作为上海近代高级公寓，其所经历的90多年坎坷历史从住户变迁中可以看出。最初的住户大都是外国驻沪公司的高级职员或具有高等教育背景的医生、教授等人士，1949-1976年这里居住着职工、政府人员和红卫兵，资本家和高级知识分子被迫迁出，如今居民背景混杂，除了少量老年常住居民外，大部分是新近住客，包括外籍人士在内的外地人士居多。居民背景的变化趋势成为历史研究和未来业态定位的基础，结合基地调研的深入展开，除对基地内其他各栋建筑的现状结构、朝向、使用情况等所作的综合调研外，还须走访黑石公寓和周边其他公寓的居民，了解他们的生活状态和公寓曾经的形态，尤其是黑石公寓只有底楼需要更新利用，而楼上都有居民居住，如何进行适应性改造而不干扰楼上住户值得细究。第二，基地的商业功能定位、业态分析和商业策划对于真实项目来说至关重要，一般会有具体的设计任务书，而此次基地所有的业态定位都需学生自己完成，也就是说，任务书要靠学生自己来完善。这样，基地未来的业态定位必须建立在充分广泛的基地周边业态调研基础上，故在设计前期就明确提出了以基地为中心、以1km和500m为半径对两个区域范围展开业态分布情况的调查工作，并组织参观历史建筑开发的商业项目，如衡山雅集、武康庭等，在对调研数据分析和整理后得出了基地未来加强文化和艺术性功能的建议。第三，基地有三栋既有建筑，其中2号楼、3号楼的结构仍坚固耐用，只要进行局部改动和外立面改造即可，黑石公寓的南向空地原为开阔气派的公寓南花园，如今则被车库和人才公寓侵占，如何利用所学建筑知识提出可落地的建筑结构改造方案以及历史花园景观的再利用思路，也是选题不容忽视的设计内容，需要深入探讨。

2 关于破题

针对今年选题的特点，破题工作尤为关键，设计前期的启发性讨论帮助学生明确设计宗旨，扩展研究视野，运用实效的调研方法，从城市功能和历史保护入手。两组的调研结果和认知感受有所差别，针对各组情况确定了设计目标和重点。同济一组的概念为"侧面的风景"，这是从基地调研中对诸多现状问题的提炼后所得出的一个普遍而强烈的特征，从基地的入口与复兴中路的错位关系开始，基地内的其他建筑都与主体建筑——黑石公寓的关系形成了一种不自然的侧面关系，而这种关系是各历史时期遗留下来的，如何面对和缓和这种畸形扭曲的建筑对位关系是形成未来设计概念的价值所在，这就触碰到城市更新中许多由于历史原因造成的场地复杂性和矛盾性，从某种层面上来说，这种关系的梳理和解决并非设计能独立完成，而是要通过开发商、设计师、政府、社区等共同合作而达成，因此今年的毕业设计希望将这样真实的基地问题展现在学生面前，引发大家去思考。经过多次讨论，最终小组围绕艺术、实验、基地三个关键词展开设计，以复兴中路的历史文化变迁为入手点，将基地改造为新时代背景下具有音乐艺术氛围、面向大众艺术活动的小型实验地，以求激活该社区的艺术活力，与对面的交响乐团音乐厅及附近的上海音乐学院所创造的区域文化氛围串接起来。同济二组的设计概念为"新海派聚落"，对基地的规

and site as keywords. The design begins with the historical and cultural changes of Fuxing Middle Road to transform the site into a small-scale experimental zone that features an artistic atmosphere and is opened to the public to carry out art activities under the background of the new times, in order to activate the artistic vitality of this community, and cater to the regional cultural atmosphere formed by the symphony orchestra concert hall across the street and the Shanghai Conservatory of Music nearby. Group 2 aimed to design the site to be "new settlements for the Shanghai style", and defined the scale, regional influence and activation points. Blackstone Apartment has a pure European-style facade, and on the side facing the street, the Corinthian columns prop up the gorgeous terrace. In the ground floor of the south-facing garden is an indoor swimming pool, which implies the luxury style of the building in its early days. With time going on, these characteristics of luxury have died down after going through many changes of the ownership, leaving nothing but the relatively intact facade. Therefore, the Shanghai-style culture and elements of the 1920s and 1930s should be integrated into the newly-created design style on the premise of breaking the closed boundary of the site, removing the traffic obstacles between the site and its surroundings such as Elizabeth Apartment and Talent Apartment, and preserving the Western style of Blackstone Apartment during the research of the modern apartments around, in a bid to create a new Shanghai-style urban settlement that can extend its influence to the surrounding communities.

3 About Cooperation

The 6 students who participated in the design project of this year had a broad range of subject backgrounds. Two of them majored in architecture, two majored in historic architecture conservation engineering, and the rest two were foreign students. They were divided into two groups of their own accord. All of them were highly motivated, but due to the difference in character, cognition and ability, they spent a long time adapting to one another in the cooperation. Upon the beginning of research at the early stage of design, the students had their traits revealed. The three members of Group 1 looked expectantly toward to the graduation design and therefore called forth all his energy. The intra-group communication and cooperation went on smoothly, but at the early stage of design, they felt confused about where to start and how to summarize site problems. Therefore, they delayed giving an interim report. However, at the late stage, they kept broadening their mind, optimizing the scheme and refining the highlights, well preparing for the final report. Group 2 was made up of a student majoring in architecture, a student majoring in historic architecture conservation engineering and a Korean student. They had their respective distinguished characteristics and always had difficulties in communicating and cooperating with each other, so a good interaction failed to be formed within the group. Despite everybody's effort and the exquisite site model made by the three members, their final achievements failed to form a joint force. As a result, the final report did not make a hit. Throughout the over three months of design, all the members made unprecedented efforts. Instead of avoiding any difficulties under pressure, they met the difficulties together with the mentor. Especially when the final report was given, they all got up the courage, and submitted the scheme to the jury confidently, leaving no stones unturned. The group leader Li Shu and the teaching assistant Cheng Cheng spared no pains to serve the group members. Their spirit is worth approving and praising.

The graduation design project was brought to a successful close at Beijing University of Civil Engineering and Architecture. In the past over three months, everybody's effort has reaped a high payoff. Despite their childish ideas, which were divorced from reality, the students strengthened their urban consciousness, historic conservation consciousness and engineering practice consciousness, saw the difference between their ideals and the reality, and recognized their own weaknesses during design. Most of all, they will always remember how they did the team work actively, took heart of grace, overcame the inferiority complex and showcased their own talent in their youthful days.

模、地区影响力和激活点作了限定。黑石公寓有着纯欧式的外观立面，沿街面上柯林斯双柱托起了气派的二层露台，南面朝向花园的底楼设置了室内游泳池，显示了当年的豪华格调。随着时间的推移，这些奢华特征在经历了产权多次易主的曲折命运后逐渐衰亡，只留下外观立面。因此，这次设计以周边近代公寓群的调研为契机，打破基地封闭的边界，疏通与周边伊丽莎白公寓、人才公寓等交通阻隔，在保存黑石公寓西式建筑风格的前提下，将20世纪二三十年代的海派文化以及时代元素一并融入到新创建的设计风格中去，力求打造一个辐射周边社区的新海派文化下的城市聚落。

3 关于合作

今年六个学生的学科背景和来历可谓丰富，其中两位建筑学、两位历史建筑保护工程，还有两位是留学生，本着自由成组的方针，最后分为两组。学生的积极性都较高，但由于学生的性格、认知和能力有所不同，合作时会有一段较长的磨合期。从设计前期的调研开始，学生的特点就显露出来，其中设计一组三个同学对此次毕业设计都充满期待，全力以赴，小组沟通和合作也顺利，不过在设计前期对设计着眼点和基地问题的归纳比较困惑，因此中期汇报时进度较慢，到了后期不断开拓思路，深化方案，凝练亮点，为最后的汇报做了充分的准备。设计二组的三个同学为建筑学、历史建筑保护工程以及韩国留学生，他们各自特色鲜明，在沟通和合作上始终有些别扭，因此小组内无法形成良好的互动关系，尽管大家都很努力，也制作了精致的场地模型，但三人的最终成果无法形成一股合力，汇报时冲击力就减弱了。回顾整个设计过程，三个多月里大家的努力和投入都超出了以往，在面临压力和困难时也没有逃避和退缩，而是和指导老师一起共同迎接和克服，尤其是最后的汇报都鼓足了勇气和信心，自信地将方案呈现在评委面前，发挥出了自己的最佳水平。其中毕业设计组长李淑一同学和教学助理程城不辞辛劳为大家付出的精神值得肯定和赞许。

今年的毕业设计最后在北京建筑大学顺利闭幕，回顾三个多月走过的路，大家所付出的辛劳都有了不小的收获和回报。尽管学生的种种设想还很稚嫩，与实际有不少距离，但设计过程让学生不仅加强了城市意识、历史保护意识和工程实践意识，也看到了理想与现实的反差，认识到自己的不足和缺陷。最重要的是，积极投入到小组的合作中来，战胜逃避和自卑心理，活出各自的精彩人生，这将成为青春岁月里最难忘的记忆。

A Pilot Study on the Early Research of Old Building Renovation Design

Xue Ying

The 2017 "Interior Design 6+1" University-Enterprise Cooperative Graduation Design project is bringing our attention to the protection and renovation of the traditional Guangdong dwelling houses in Guangdong. Our design project is located in the old urban area of Guangzhou—in the southeast of the planned red line of the commercial and residential building in Zhuguang Road, Yuexiu District, bordered by the former Overseas Chinese Merchant Street. Now it has been acquired by Yuehai Real Estate Limited and will be transformed protectively to meet the requirements of modern production and living.

1 The Significance of Early Research

A new function of architecture is to solve the first problems arising during renovation design, and this is a prerequisite to design. We are trying to apply rational research methods to the graduation design and solve the problem of function positioning through research. The main aim of research is to sort out region information to identify all problems, explore the historical cultural value of architecture, and seek new social, economic and environmental values. Therefore, research has become a fundamental basis for the new function positioning of the traditional dwellings, as well as a key part and important guarantee of preliminary design, and the result of research will provide support for design. In traditional teaching, the importance and objectivity of research are often neglected, but this time, we hope the design research can go beyond the traditional teaching, which advocates describing phenomena subjectively, and instruct students in relinquishing their subjective thought to identify, analyze and solve problems.

The research of old building renovation design differs from that of general building design, because renovation design is made on the basis of protecting and respecting the historic content of an old building. This dwelling house has a high historic value. The combination of the traditional terraced bamboo tube-shaped rooms and the Western-style arched red tile facade makes it a unique historic building in the old urban area of Guangzhou, but the building has been submerged in the tall buildings around and does not attract attention from the community residents. So, we wish to research its history to evoke the memory of it.

The dwelling house is located in a residence community, and the surrounding environment gives it a strong social attribute. We are trying to continue the historical spirit in this old building, and meanwhile open it to the public, because after renovation, it will be used and managed by the public. The design is made based on a huge user group, and this means that we should have in mind the needs of potential users, so the community users are key respondents. Users need to carry out activities in the old building, and the activity content concerns the functional layout and the economic value of the old building, thus the surrounding industries are also important respondents.

In the course of research, the students not only were designers, but also played various temporary roles in going deep into the community in a bid to find out the benefit relationships among different groups and the relations between economic benefits and social benefits. In a word, to fully explore the historic value, social value, economic value and environmental value of this traditional dwelling house, we should not limit our work to design. Instead we ought to do research into the social cultural phenomena behind design. Design is in itself a factor of social culture, thus we hoped that the students could personally feel the social issues and gain the firsthand information of design through research, and that the result of research would provide support for design.

2 Several Methods applied to Research

Under the teachers' guidance, the students could, do research into the complex issues of the communities around the old building by SWOT induction, tracking, interview and questionnaire survey. At this stage, the students must have keen observation ability, analytical ability, graphic expressive ability and communicative ability.

2.1 SWOT Analysis

Before field research was done, the students could look up document literature by all means, to gain a preliminary understanding of the project's profile and circumjacent information. The document content was usually all-embracing, and in order that the students could find right data in massive information, we adopted the SWOT analysis method, an approach used in the commercial world. SWOT is a strategic planning and analysis tool commonly used in the business domain. The English letters stand for strength, weakness, opportunity and theatre respectively. This method is to make a comprehensive analysis of the internality and externality of an enterprise, including its strengths, weaknesses, opportunities and threats.

First, the students made an inductive analysis of the document literature to summarize the site and building's strengths, weaknesses, opportunities and threats from the perspective of architecture, community condition, community residents' basic information, site environment and historical cultural environment. The SWOT method also applies to the consolidation of field research information. The content of field research is complex, but with the guidance of a SWOT analysis sheet, the students will pay attention to the building's strengths and weaknesses, and summarize complicated information. This method can help to make best use of the advantages and bypass the disadvantages, reveal the strengths of this traditional dwelling building and make it adaptive to the modern production and life. For a student without too much experience in research, the SWOT method is really a good analysis method. It covers a wide range, and the chart is straightforward, readable and directional, so it can help students to do more than simply list research information.

Of course, the SWOT is not a one-fit-all analysis method, since it just presents objective real material information, while it does not judge or collate invisible information, such as people's personalized information. Therefore, we applied the tracking

旧建筑改造设计课题中的前期调研初探

薛颖

2017年"室内设计6+1"校企联合毕业设计的课题引领我们关注广东传统民居的保护与更新问题。我们的设计项目位于广州老城区——越秀区珠光路商住楼规划红线东南角，毗邻原侨商街。这栋三层高的建筑建于1930年代，被列为广州市第一批历史建筑，总建筑面积1852.30 ㎡，现被粤海地产公司征用，它将被保护性地改造，以适应现代生产、生活需要。

1 前期调研的意义

建筑新功能的定位是改造更新设计过程中首先面临的问题，也是设计的前提。我们试图将较为理性的调研方法引入这个毕业设计课题，通过调研解决设计功能定位问题。调研的主要目的是梳理场地信息以明确存在的问题，发掘建筑的历史人文价值、寻求新的社会、经济、环境价值。因此调研成为传统民居新功能定位的基本依据、设计前期的关键环节和重要保障，调研的成果将成为设计的依托。以往的教学容易忽视调研的重要性及客观性，希望这次设计调研能突破传统教学中对现象主观描述的思路，带领学生从自己的主观臆想到依据事实发现问题、分析问题、解决问题这一过程的实现。

旧建筑改造设计的调研与一般建筑设计调研有所不同，因为改造设计是基于保护、尊重旧建筑的历史文脉，在此基础上进行更新。这栋民居非常具有历史价值，传统联排竹筒屋的建筑形制与西式连续拱券的红砖立面结合，使得它成为广州旧城独具一格的历史遗存，但是这栋建筑被众楼房淹没，并未引起社区居民的注意，因此我们希望通过研究它的过去，让历史重新被唤醒。

这栋民居坐落于住宅小区，周边的环境赋予它很强的社会属性。我们试图使旧建筑延续历史精神的同时，还要使这栋旧建筑走向开放性与公众性，因为改造后的建筑使用者和管理者是社会群体。设计是建立在广泛的使用者的基础之上，这意味着我们要把握潜在使用者的需求，因此社区用户群体成为调研的重要对象。用户在旧建筑里需要展开活动，活动的内容关系到功能布局，关系到旧建筑的经济价值等问题，因此对于社区周边相关业态的调研也是我们重要的调研工作内容。

调研过程中，学生不仅是设计师，还临时扮演各种角色深入社区，企图发现不同群体的利益关系，发掘经济利益与社会利益之间的关系。总之，要充分发挥这栋传统民居建筑的历史价值、社会价值、经济价值、环境价值，并不是单纯关注设计问题，而是需要通过调研了解设计背后的社会文化现象。设计本身就是社会文化的一个因子，我们希望通过调研，学生能真切感受社会问题，确凿掌握设计的第一手资料，也希望调研的成果最终反哺于设计。

2 调研过程中运用的几种方法

在老师们的指导下，同学们运用SWOT归纳法、跟踪法、访谈法、问卷调查法等方法对旧建筑周边社区的复杂问题进行调查研究。在该阶段，学生必须具备敏锐的观察能力、分析能力以及图解表达能力、交流能力。

2.1 SWOT归纳分析法

进行实地调研之前，学生利用各种途径进行文献资料的查阅。对项目周边情况及项目概况形成初步了解。资料文献的内容往往是包罗万象的，为了让学生从众多的信息中聚焦调研目的，我们引用了商业界SWOT的分析方法。SWOT归纳分析法是商业领域常用的战略规划分析工具，英文字母分别代表企业优势（strengths）、劣势（weakness）、机会（opportunity）和威胁（threats），SWOT的分析是将企业内部和外部各个方面进行归纳，分析企业面临的优势、劣势、机会和威胁的一个方法。

学生先将文献资料进行归纳，从建筑单体、区位交通、社区业态、社区人群基本情况、基地环境、基地历史人文环境等方面总结出基地和建筑自身的优势、劣势、机会、威胁。SWOT方法同样适用于现场调研信息的整理，现场调研的内容繁杂，有了SWOT分析表的指引，学生便会关注建筑的优势和劣势，对复杂的信息进行归纳总结，旨在扬长避短，挖掘这栋传统民居建筑的优势并适应现代人的生产和生活。对于没有太多调研经验的学生，SWOT分析法不失为一种不错的分析方法。它涵盖面广，图表结合的方式直观、宜读，指向研究目的，一定程度避免了学生对于调研信息简单的罗列。

当然SWOT不是万能的分析法，SWOT只呈现了客观的硬性的物质信息，并未对无形的信息，例如对于人群的个性化信息进行判断和归纳整理。因此，对于人群的个体研究我们采用了跟踪法、访谈法、问卷调查法。

method, interview method and questionnaire survey to the individual research.

2.2 Tracking, Interview and Questionnaire Survey

The function and quality of an interior space depend on the closeness of the combination of the environment and human behavior, as well as the changes of the behavior pattern in the environment, thus the research of the user activity system is of great importance. Since this traditional dwelling house was still vacant, wen could only do research into the community residents and possible users. The concrete content of research was designed by the students.

What makes tracking, interview and questionnaire survey the most different from SWOT is that they are used for different research objectives. The first three methods try to start from human behavior to explore the problems behind speeches in subtle acts and casual words. This is what we call "making a fuss over a trifle". However, these methods are hard to apply since they rack brains, and that all social issues identified by these methods are not effective information or solved as design issues.

Some students observed the surroundings of the site statically and dynamically from 6 o'clock to dawn. The students saw how the citizens were leading a civil life. They found a temporary market, on which trade activities were just held at dawn and most goods were collectibles. Since urban management officers were not on duty at dawn, the market was very active at that time. Then, this information provided an inspiration for the outdoor environment design of the old building.

The students learned from the respondents that the community residents were not sensitive but indifferent to this building which has a historic value. This phenomenon reveals the lack of the community spirit. The students learned form the questionnaires that the community residents were badly in need of greening and a public space. This information has become a basis for interior function positioning.

The content of the questionnaire needed to be designed. First, the designer should know the purpose of questionnaire survey, so as not to waste or omit important issues. Second, issues should be concrete rather than abstract. Issues should be popular and easy to understand and sorted from easy to difficult.

2.3 Participative Design and Temporary Role-playing

When we renovate the old building, we should make known all different memories of the building to ordinary citizens, and this will reflect the importance of social participation. Participative design stresses that space users should be empowered during design so as to intervene in spatial environmental decision-making. Empowerment refers to that different social roles and groups are empowered to actively participate in the process of design to meet their own requirements for a specific space.

It is a new attempt to introduce the rational spirit represented by participative design into the teaching domain of environment design. Participative design was simulated during the research, in which we analyzed users' activity and behavior styles, defined the purpose and task of construction, and finally implemented the purpose and task during design. Although participative design is still in its infancy in Chinese mainland, and the method cannot be really applied to a graduation design project, we can use some research methods for reference.

Participative design pattern is an open system, in which all parties concerned can take a part to put forward their requirements and viewpoints, and then the designer can serve as the organizer to complete the practical operation. Since it fully respects users' right, consider users' actual demands, participative design calls for user participation upon project positioning to show the designer users' intention and goal, so that the designer and users could jointly look for an optimal way to meet the demands. Such sustainable development values that emphasize public participation in spatial place design, individual intervention design and reuse of old buildings are an organic unity.

The students played various roles in the course of design. They sometimes played a designer, sometimes a developer, and sometimes a user, community resident or government officer, in an attempt to understand the relationship between the community residents and the developer, the relationship between the government and the developer, and the relationship between the designer, developer and community residents. By playing temporary roles, the students thought on the standpoint of different groups and simulated different groups' needs in the hope of harmonizing various conflicts by design, so as to safeguard most people's interests to the greatest extent.

When simulating participative design and playing temporary roles, the students indeed came up with many design ideas and schemes, which might not be able to help develop a preferred solution, the students learned how to focus their attention on social issues and reflect upon the designer' social values, historical values and mission of the times. This goal should become an important part of design teaching.

3 Conclusions

The research of graduation design has enlightened us as follows: there is no completely objective knowledge in design, but every problem must be analyzed rationally. The results show that none of the above methods can ensure absolute science or objectivity. Graduation design research is scarcely independent of a subjective evaluation, which is neither absolutely right nor absolutely wrong, so all that matters most is a rigorous thinking process. Design issue is a social cultural issue, which needn't be treated absolutely objectively, but most of all, it is necessary to identify problems, understand the society and set up positive values through research, so as to finally solve the problems through design.

2.2 跟踪法、访谈法、问卷调查法

室内空间的功能及品质取决于环境与人的行为结合的紧密程度以及环境中行为模式的变化，因此对于使用者活动系统的调研非常重要。由于这栋传统民居建筑处于空置状态，我们只能对社区居民及将来可能的使用者进行调查研究，具体的调查内容由学生设计。

跟踪法、访谈法、问卷调查法与SWOT分析法最大的差别在于调研目标不同，前三种方法企图从人的行为出发，从细微的举动、不经意的言语中探究行为言语后面的问题，可谓是"小题大做"。但是运用这些方法颇费心机，发现的社会问题并非都是有效信息，并非能转换为设计问题。

学生曾经从清晨6点到凌晨，对基地周边进行蹲点观察和跟踪观察。学生观察到的是非常市民的生活情景，他们曾发现了一个虚市，它只在凌晨的街道上产生交易活动，交易的物品主要是卖家收藏的物品，由于凌晨没有城管的干预，这个虚市一直活跃于凌晨时分。因此，这个信息成为旧建筑室外环境设计的一个灵感来源。

学生通过访谈，得知社区居民对于这栋有具历史价值的建筑其实并不敏感，也漠不关心，这种现象让我们感到社区精神的缺失。学生通过问卷方式得知社区居民最需要绿化和公共活动的场所，这些信息成为室内功能定位的依据。问卷内容是需要设计的，首先，问卷设计者要明确问卷的目的，避免问题浪费或是遗漏重要问题；其次，问题应具体而不是抽象。问题应该通俗易懂，按照先易后难的前后顺序排列。

2.3 参与式设计和临时角色扮演方法

当我们对旧建筑进行改造时，建筑所负有的不同层面的记忆应该可以被普通市民所认知，这就体现出了社会参与的重要性。参与式设计强调空间使用者在设计过程中被赋权，真正影响空间环境的决策。赋权指的是赋予社会不同角色和群体应有的发展权，他们能够主动地参与到设计的过程中，满足自己对具体空间的要求。

参与式设计代表理性的精神被引入环境设计专业的教学领域是新的尝试。此次调研过程模拟了参与式设计的做法，分析使用者的活动与行为方式，明确建造的目的与任务，最后将目的与任务付责设计实施。虽然参与式设计的方法在中国大陆还处于摸索阶段，而且对于毕业设计并未能真正实施这种方法，但我们可以借鉴参与式设计的一些研究方法。

参与式设计模式是一个开放的系统，相关各方参与进来，提出自己的要求与看法，由设计者承担组织者的角色，完成实际操作。参与式设计充分尊重使用者的权利，考虑使用者的实际需求，从项目定位初始就让他们参与进来，及早让设计者了解使用者的意图和目标，然后设计者和使用者共同寻找最佳途径去满足需求。这种强调空间场所设计的公众参与性、注重社区个体介入设计与旧建筑再利用的可持续发展价值观形成了有机的统一。

学生在模拟参与式设计过程中扮演各种角色，有时是设计师，有时是开发商，有时是使用者、社区居民、政府人员，试图理解社区居民与房地产商的利益关系，政府与房地产商的关系，设计师与房地产商、社区居民的关系。学生们通过临时角色扮演的方法，站在不同群体的立场换位思考，模拟不同群体各自的需求，希望通过设计手段协调各种矛盾，最大程度保证多数人的利益。

通过模拟参与式设计和临时角色扮演方法的结合，学生确实产生了多种设计思路、设计方案，也许未必能得到最佳的解决方案，但是却促进学生关注社会问题，思考设计师的社会价值观、历史价值观和时代使命。这个目的应该成为设计教学中很重要的一个内容。

3 结语

这次毕业设计的调研活动对我们的启发是：设计中没有纯粹客观的知识，但是却不能没有理性地分析问题的过程。结果发现，无论使用哪一种方法，我们都无法做到绝对的科学、客观。学生毕业设计课题的调研难以脱离主观的评价，而主观的评价并没有绝对正确或是错误，重要的是要具备较为严谨的思辨过程。设计问题是个社会文化问题，并不需要绝对的客观，重要的是能够通过调研发现问题，认识社会，树立积极的价值观，最后再通过设计的手段应对问题。

A Reflection upon the Technical Concept Education in the Graduation Design

Zhou Lijun, Ma Hui, Zhao Hui

Abstract: The graduation design is used to test a student's comprehensive capability. In view of the objective of professional designer cultivation, the paper emphasizes the implementation of relevant technical concepts in the graduation design, and illustrates the important role of technical concept education in the graduation design by analyzing the application of structure technology, energy saving technology and digitalization technology in the graduation design.

Keywords: graduation design, technical concept, design expression

When admitted into the school, students have had such a common sense: architecture is a subject composed of technology and art, which are supplementary to each other and indispensable. But compared with "invisible" technology, students seem more concerned with the form and style. In their course sheet there is no lack of architectural physics, architectural structure, building materials or other technology courses, but since a teacher generally just teaches a certain course in practical teaching, the inter-course relation is very poor, and students have many difficulties in making a thorough master of the knowledge, especially in applying the architectural technological knowledge to architectural design. However, the graduation design can just make up for the deficiency, since it stresses comprehensive use of the relevant basic theoretical knowledge, professional knowledge and basic skills to analyze and solve practical problems.

1 The Target and Features of Graduation Design

The teaching process of graduation design is a stage important for the realization of the undergraduate training objective; in the syllabus of this specialty blocked out by the Architecture Specialty Committee, graduation design is positioned as: a comprehensive summary of the five-year undergraduate teaching, used to test students' comprehensive ability. The graduation design in architecture offered at Harbin Institute of Technology focuses on developing and testing students' creative spirit and practical ability. In many years of teaching practice and exploration, the School has formed a characteristic that technical concepts are expressed particularly in the teaching of graduation process.

Normally, the graduation design has a long cycle and much content, which requires students to make a profound project by applying all the knowledge learned in the first four years to the project, so it is quite complicated to design a title. Also, the project would better be done based on a real production or scientific research task. All the "Interior Design 6+1" school-enterprise cooperative graduation design projects hosted by the Institute of Interior Design of Architectural Society of China are real projects and done based on complete basic materials such as topographic maps, planning opinion books and assignment books. In such a practical environment, students may have their learning interest enhanced, familiarize themselves with the operational sequence of practical projects, and gain comprehensive practical experience, so as to transform their knowledge into practical ability to the utmost extent before graduation, to grow into a professional architect as quickly as possible.

2 Technical Concept Analyses

"Technology was first defined by Diderot in Encyclopedia as 'various tools and rule systems called together for a certain purpose.' Later, with the development of technology, its role expanded and the range of its definition further extended. In the modern literature, there are various new definitions of technology, such as 'technology is the sum of the means of human activities', 'technology is the materialization of science', etc."[1] It is self-evident that technology is of great significance to architecture, there is an inseparable inherent relationship between architecture and technology, and the development and breakthrough of technology has greatly promoted the development of architecture throughout the history of architecture. China's overall architectural creation level is far lower than the developed countries', and one of the main causes is that Chinese architects are less capable of deeply understanding and practicing technology. The reflection shows that the originality of a design scheme and the beauty of a picture cannot be highlighted alone, while the technical concept expression should be enhanced.

For the technical concept expression, students needn't have to fully understand how to apply every technique in detail when doing a project, but need to fully understand technology's great significance to architectural design; students are not required to master any sophisticated architectural technology, but they should be able to select a suitable technical route in accordance with the actual situation, and lay stress on the combination of technology with natural ecology and cultural texture with a specific economic condition.

3 Technical Concept Expressions

As a "materialized invisibility", technology's expression in the graduation design can be reflected in many aspects such as structure technology, energy saving and low carbon technology, architectural physics technology and design technology.

3.1 Structure Technology

In the title of a graduation design project, a technique that is mostly applied is structure technology. Especially for the title of a project related to the design of a long-span public building or the transformation and renovation of an old building, students study more problems related to structure technology. They need to carefully solve every problem related to structure selection, mechanical analysis and construction mode design. Of course, acoustic technology is also essential for theatrical buildings.

3.2 Energy Saving and Low Carbon Technology

Nowadays, energy saving and low carbon technology is really a focus of attention in the architecture circles, and school

毕业设计中技术理念教育的思考

周立军 马辉 兆翚

摘要： 毕业设计是对学生综合能力的考察，针对职业设计师的培养目标，文章强调相关技术理念在毕业设计中的贯彻实施，并通过分析毕业设计中结构技术、节能技术、数字化技术等的应用，阐释技术理念教育在毕业设计中的重要作用。

关键词： 毕业设计 技术理念 设计表达

学生从入校伊始就知道这样一个基本常识：建筑学是技术与艺术相结合的一门学科，二者相辅相成、缺一不可。但是与"隐性"存在的技术相比，学生们似乎更关注形式与风格。在他们的选课单中不乏建筑物理、建筑结构、建筑材料等技术性课程，然而在实际教学中，一个任课教师一般只是讲授其中的某一门课程，导致课程之间横向联系较差，学生很难将所学的知识融会贯通，在建筑设计中不知怎样去运用建筑技术知识。毕业设计的设置恰恰可以弥补这方面的不足，它强调综合运用所学基础理论知识、专业知识和基本技能来分析和解决实际问题。

1 毕业设计的目标与特征

毕业设计教学过程是实现本科培养目标的重要阶段，建筑学专业指导委员会为本专业拟定的大纲中将毕业设计定位为：毕业设计的目标是对学生综合能力的考察，是整个五年本科教学过程的一次综合性的总结，哈尔滨工业大学建筑学专业毕业设计的教学注重对学生的创新精神和实践能力的培养与考察。在多年的教学实践和探索中，我院形成了注重毕业设计中的技术理念的表达的教学特点。

毕业设计的设计周期一般较长，同时设计内容较多，要求学生的设计要达到一定的深度，将前四年中所学的知识在毕业设计中综合运用，因此设计题目通常具有相当的复杂性。中国建筑学会室内设计分会主办的"室内设计6+1"校企联合毕业设计都是以真实项目作为毕业设计选题，地形图、规划意见书、任务书等基础材料齐全，使学生在实战氛围的感染下，提升学习的兴趣，熟悉实际工程的操作顺序，获得较全面的实战经验，促使学生在毕业前最大限度地完成从学习到实践的转变，尽快地实现从学生向职业建筑师的蜕变。

2 技术理念解析

"技术最初是由狄德罗在《百科全书》中所下的定义'它是为某一目的共同协作组成的各种工具和规则体系。'后来随着技术的发展，其作用不断扩大定义的范围也进一步拓展。在现代文献中出现了各种新的技术定义，如'技术是人类活动手段的总和''技术是科学的物化'等等。"[1] 技术对于建筑的重要意义是不言而喻的，建筑与技术有着密不可分的天然联系，技术的发展和突破对建筑巨大的促进作用贯穿建筑史的始终。中国的建筑创作水平总体上远落后于发达国家的主要根源之一就是建筑师缺乏对于技术的深入了解和付诸实践的能力。在反思中我们发现，不能仅仅偏重于设计方案的独创性与图面表达的赏心悦目，而是应该强化技术理念的表达。

在学生的设计过程中，对于技术理念的表达并不是要求学生完全深入地了解建筑中每一项技术的具体应用，而是充分认识到技术对于建筑设计的重要意义；并不是要求学生一定要掌握多么高深的建筑技术，而是能根据实际情况选择适宜的技术路线，注重技术与自然生态、文化肌理和具体经济条件的结合。

3 技术理念的表达

作为一种"物化形态"，技术在毕业设计中的表达可以体现在结构技术、节能低碳技术、建筑物理技术以及设计技术等诸多方面。

3.1 结构技术

在毕业设计的题目中，较多应用到的技术就是结构技术，尤其对于大空间公共建筑、旧建筑改造与更新的题目来讲，学生需要较多的涉猎有关结构技术方面的问题。从结构选型到力学分析和构造方式设计，都是需要认真解决的。当然，在观演建筑中声学技术等也是必不可少的。

3.2 节能低碳技术

当今，低碳节能技术可谓建筑界所关注的焦点，而学校的教育原本应该走在行业前面。目前，对低碳的关注有一些流于表面，把节能作为一种时髦字眼装饰设计作品的做法不在少数，而勇于挑战技术难题则需要相当的技术创新。

education on it should have been in the front rank. At present, the attention to low carbon technology is formalistic in a sort of sense, and it is not unusual that energy saving is taken as a vogue and used to decorate design works, while hi-tech innovations have to be made for a challenge to technical problems. It is also necessary to perform energy saving technology software simulation and analyze and solve some technical problems about design by quantitative measures.

3.3 Design Performance Technology

The contemporary college students, who grew up in the digital age, have a lifestyle closely related to electronic products. Although we keep encouraging students to enhance the most primitive freehand design, we have no way to reverse the tide of the digital revolution. At present, the award-winning works in lots of large-sized international competitions feature the concept and technology of complicated design. Therefore, it is very important to learn and correctly use digital performance technology.

4 Conclusions

The aim of architecture education is to produce all-round professionals demanded by the society, and graduation design is a turning point in the whole educational process, while the implementation of a technical concept is an important condition for this point to be put into practice. Besides, the expansibility and diversity of technology also decide the dynamism of the expression of such a concept. Whether it's to apply structure technology, protect environment in a low-carbon energy-saving way, a apply architectural physics technology in a concrete manner or usher in a new trend of architectural design technology, it requires that the exploration on the reform of graduation design should have the characteristics of the times, and call for continuous reforms, innovations and breakthroughs with the rapid development of science and technology, to make graduation design teaching more forward-looking and developmental.

References

[1] Sun Cheng, Mei Hongyuan. The Technical Concept in Modern Architectural Creation[M].Beijing: China Architecture & Building Press, 2007.

还需要掌握和选择运用节能技术软件模拟,通过量化的手段分析和解决设计的一些技术问题。

3.3 设计表现技术

目前的在校大学生是在数码时代成长起来的,他们的生活方式与电子产品息息相关。我们虽然一再鼓励学生们强化最原始的徒手设计方式,但是却无法逆转数字革命的浪潮。目前很多大型国际竞赛的获奖作品,都得益于复杂性设计的理念和技术。因此数字表现技术的学习与正确应用是十分重要的。

4 结语

建筑教育的目的是培养适应社会的全面发展的专业人才,而毕业设计则是这一整套教育环节中具有转折性的节点,技术理念的贯彻是这一转折得以实施的重要条件。同时技术的发展性和多样性的特点也决定了这种理念的表达是动态的,无论是结构技术的应用,还是低碳节能环境保护,或是建筑物理的具体技术应用,亦或是迎接建筑设计技术的新趋势,毕业设计的改革探索应具有时代特征,应随着科学技术的快速发展和生产方式的变化不断变革、创新和有所突破,使毕业设计的教学更具有前瞻性和发展性。

参考文献

[1] 孙澄,梅洪元. 现代建筑创作中的技术理念 [M]. 北京:中国建筑工业出版社,2007.

A Reflection on Teaching provoked by "Conservation and Utilization Design of Traditional Dwelling Houses"

Gu Qiulin, Liu Xiaojun

Introduction

The Architectural Society of China's 5th (2017) "Interior Design 6+1" school-enterprise cooperative graduation design title, which is known as "Conservation and Utilization Design of Traditional Dwelling Houses", is of great contemporary significance. Based on the title, every university joined hands with its cooperative enterprise and selected a specific heading according to a classical local dwelling building. Finally, "7 participants, as well as their 7 partners, selected 7 topics". Our students selected the traditional cave dwelling architecture which has the distinctive feature of northwest residential buildings.

This topic matters much to the protection of architectural heritages, to which various countries have paid high attention. As urbanization advances and urban population increases sharply, the traditional architecture pattern and volume are already far from being able to meet the current demands. A great many traditional dwelling buildings have been demolished and replaced by residential buildings and commercial complexes. At present, few traditional dwelling buildings are left, namely few available resources left. How to better protect and use the only resources? It is an issue that should be thought about by the whole society today, and much of an issue that must be researched by the contemporary designers.

For an undergraduate, it is not easy to research this topic, because it is still a hard nut to crack for many of the contemporary experts. For students, they don't know where to start; thus as instructors, we should first help them to analyze and summarize the following problems: Is reconstruction required? What is the basis? How should reconstruction be performed? What methods can be applied? What is the ultimate goal of reconstruction? Then, we will help them at every stage by organizing teaching rationally and reforming the teaching method.

1 Taking early research seriously to foster the observation ability and data acquisition ability

Research at the early stage of design is the precondition and foundation for students' design, and a task that must be done at the early stage of design. Due to their insufficient understanding of the importance of early research, as well as the relationship between the early research and post-design, many students often belittle or ignore this step, and just collect insufficient data. As a result, they may fail to accomplish the design or achieve an ideal result. Therefore, we should impress early research on students, to mobilize their subjective initiative, and foster their ability to observe, research, record and collect data. We spent at least 2 weeks doing social research. We asked the students to go out of the campus to investigate the project sites, and make an effective analysis, comparison and evaluation on the design themes, design backgrounds, and the sites' natural and cultural factors. The research was basically conducted in the following steps:

(1) Clearly analyze the assignment book to find out something interesting and elusive.

(2) First look up online data, such as theses, or relevant cases.

(3) Do research before field verification, visit the site to feel it, and make a further comparison through questionnaire survey, interview, photographing, and human behavior analysis. The above work may need to be done several times, since omissions are inevitable.

(4) First make a mind map, and then make a drawing or PPT by categories. Be sure to keep the logic clear.

(5) Then query cases, but query shouldn't be limited to the title. Try to read as many cases as possible to spark inspiration.

In this way, the students can get an all-around understanding of the design content, so as to lay a solid foundation for the in-depth design at the later stage.

2 Focusing on fostering students' creative ability

As is known to all, the level of students' creative ability directly decides the success or failure of a scheme. In school education, students' creative power is shown in the collective, so their creativity cannot be fostered without a free, safe and pleasant collective atmosphere. To create a creative collective atmosphere, teachers should first admit students' developable great creative potential and then provide them an adequate opportunity, reduce the unnecessary restrictions and regulations, and ask them to study creatively or join other creative activities.

Secondly, creative teaching is necessary. Creative teaching is a way of teaching carried out by a scientific method and approach under the guidance of the basic principles of creative studies, creative psychology and creative pedagogy in order to not only impart knowledge and develop ability, but also foster creativity and explore creativity. In laymen's terms, teachers must have problematic consciousness from beginning to end, since problematic consciousness is the basis of creativity cultivation. For design guidance, we should start with the teaching philosophy, teaching atmosphere and teaching process to guide teaching in a creative way.

(1) In terms of teaching philosophy, we should turn the students' attention from "learning" to "asking", and get our teeth into and think about every question raised by the students, so as to enable them to fully display their own thought and personality.

(2) In terms of teaching atmosphere, we should awaken the students' problematic consciousness, and enhance their motivation of raising questions. We may as well give them an opportunity to express their ideas freely and identify problems, make favorable comments on their answers, and inspire them to look for a correct design idea rather than tell them the answer in an oversimplified and crude way.

(3) In terms of teaching process, we should actively foster their problematic consciousness to improve their ability to raise questions, because question-posing can help to inspire them to develop more solutions.

由"传统民居保护利用设计"引发的教学思考

谷秋琳　刘晓军

引言

本届中国建筑学会室内设计分会"室内设计6+1"2017（第五届）校企联合毕业设计的选题非常具有时代意义，题目为"传统民居保护利用设计"。本次选题的具体方向，由参加高校分别联合各自相关企业，从所在区域典型传统民居中选定，形成"7所参与高校+7家合作企业+7类民居课题"的新形式。我们的学生则选择了最具有西北民居特色的传统窑洞建筑。

该课题切中了当下各个国家都非常关注和重视建筑遗产的保护问题。随着城市化的不断加剧，城市人口急剧增多，原有传统建筑格局及体量已远远不能满足现在的使用需求。大量的传统民居建筑被推倒，新建为居民楼与商业综合体等，传统民居建筑所剩无几，可用的资源已所剩无几。如何更好地保护与利用这些仅存的资源？这是今天全社会都将思考的问题，更是当今的设计师必须研究的问题。

对于本科毕业生来说，此课题的研究还是具有一定难度和深度的，因为这毕竟是我们当今众多专家学者还在探讨的问题。同学们刚开始都不知该从何入手，因此，在指导学生设计之初，就先要帮助他们分析和归纳出以下几个问题：要不要改造，改造的依据是什么？如何进行改造，改造的手法有哪些？怎么改造，改造的方向在哪里？然后，再通过合理的教学组织、教研方法的改革，来帮助他们一步步完成。

1 注重前期的调研，培养观察能力和资料收集能力

设计前期的调研是学生进行设计的前提和基础，是设计前期需要做的必要工作，很多学生由于对前期调研的重要性和调研工作与后期设计之间的关系理解不足，往往会不重视或忽略这一步骤，只是草率地收集一些并不详尽的项目资料，就会造成设计工作进行不下去或最终的设计成果不理想。因此，我们应该在前期让学生对于调研的工作重视起来，让他们发挥主观能动性，培养他们的观察能力、调研、记录和资料收集能力。我们利用了至少两周时间进行社会调研，让同学们走出校园，到实际项目基地去考察，对设计主题、设计背景以及基地的自然、人文因素进行有效的分析、对比和评估等。基本按以下步骤进行：

（1）审清任务书，找到自己感兴趣和不懂的地方。

（2）先查网上的资料，可以是论文，或者去相关网站查案例，尽量切题一些。

（3）去实地验证之前的调查，并且感受现场，通过问卷调查、走访、拍照、人的行为分析等进行深一步的比对。有时候可能要去好几次，因为难免有漏掉的点。

（4）整理成思维导图之后，再分门别类的形成图纸或者PPT，这时候，一定要层级清楚，逻辑很重要。

（5）之后再查案例的时候，就可以不局限于题目，可以多看看，能激发灵感就好。

这样以来，同学们就可以对所设计的内容有一个全方位的理解，为后期的深入设计打下结实的基础。

2 注重学生创造能力的培养

众所周知，学生们创造能力的好坏，会直接影响到一个方案的成败。在学校教育中，学生的创造力量是在集体中表现出来的，没有自由、安全、愉悦的集体氛围便不能培养学生的创造力。要营造创造性的集体气氛，教师就要在承认学生具有可以开发的巨大创造性潜能的基础上，为其提供充分的机会，减少一些不必要的限制和规定，让他们能进行创造性的学习或其他活动。

另外，还要进行创造性教学。创造性教学是以创造学、创造心里学和创造教育学的基本原理为指导，运用科学的教学方法和教学途径，在传授知识、发展能力的同时培养创造性、开发创造力的教学。简单来说就是教师必须始终要有问题意识，问题意识是创造力培养的基础。在设计指导过程中，我们应该从教学理念、教学气氛和教学过程等方面入手进行创造性的教学指引。

（1）教学理念上要从注重"学"转变为关注"问"，认真对待和思考学生提出的问题，让学生充分展现自己的思想和个性。

（2）教学气氛上要激活问题意识，增强学生提出问题的动机。给学生一个自由表达思想和发现问题的机会，以及对其问题答案的多样性的肯定，并启发他们寻找正确的设计思路，而不是简单粗暴地告诉他们这个应该怎么做才是对的。

（3）教学过程上要积极培养问题意识，提高学生提出问题的能力，有问题，才能激发出更多解决问题的办法。

在这个过程中，我们培养学生主动地学习，延迟判断，给他们足够的时间去创造；发展学生思维的灵活性；训练学生的感觉敏锐性；尽可能创造多种条件，让学生接触不同的设计理念、观点等。

In this process, we can teach them how to learn actively and make deferred judgments, and give them enough time to create; improve the flexibility of their thinking; improve the sensitivity of their feeling; try every means to create as many conditions as possible to throw them in contact with different design concepts and ideas.

3 Focusing on school-enterprise cooperation to enhance the students' ability to control and operate a project

Since Group 1 chose an actual PPP project, it mattered much whether the final scheme could be well implemented. Therefore, everybody should learn how to seek a balance between creativity and the practical operation. Additionally, the topic requires that we should turn our face to a traditional dwelling around us, participate in its construction and consider its future development. The ultimate aim of teaching is to provide a service for the local economic and social development and this will also offer us the largest class and a platform where we can showcase our talent.

Therefore, the enterprise closely related to the present topic, as a partner of our school, is a reliable guarantee for accomplishing personnel training and teaching implementation. The "Interior Design 6+1" school-enterprise cooperative graduation design project is a typical successful case, and the form of this year is more effective than before. At the present off-campus cooperation event, the representative of the cooperative enterprise will offer guidance to design throughout the event. Since he is familiar with the students' basic condition, he will have smooth communication with them. In the process, the representative will timely correct the students' fanciful ideas and design contents, tell them what is good and what is fantastic or unable to bring about any practical value, and show them new materials and technologies in teaching, to lay a foundation for their future career.

4 Focusing on fostering the students' logical thinking ability and expressive ability

Many art college students have a disadvantage that they have strong perceptual thinking and divergent thinking abilities but weak logical thinking and analytical abilities. There is no exception to our students, who usually just pay attention to the shape rather than think about the original intention of design or the problems that should be solved. They lack the ability to control design on the whole.

Moreover, due to their poor logical thinking ability, many students don't know how to make their design a methodic whole after developing a scheme. The promiscuous logical relation often confuses outsiders, and this may directly affect the dissemination of project information. Besides, most students with poor rational thinking ability have weak expressive ability. By expressive ability here is meant not only the expressive ability of design, but also the expressive ability of language, namely communicative competence. Even if a design is very excellent, if the designer cannot describe it well to explain it clearly to listeners, it will be an abortive scheme. The expressive ability of language is a skill essential for a designer.

Therefore, at the early stage of design, we must focus on fostering the students' rational analysis and logical thinking abilities, so that they could turn their perceptual cognition gradually into a rational analysis, to keep identifying and solving problems. After repeated training, they will be able to directly link the early content with the design objective, to focus on the key points at one stroke. Furthermore, during training they should learn to describe the design process step by step to identify problems, raise questions, research the problems and questions, and develop solutions, to deepen the design and reversely demonstrate the design scheme through analysis and comparison, so as to finalize the best the design scheme. In this process, we should also keep encouraging them to verbally describe their scheme, so that they could not only make a design, but also talk about it, and learn to express the scheme in the best way through constant practice, to make it understandable.

Our students were impressed the most deeply this time, and every report revealed a obvious improvement in their work. At first, they had their heart in their boots when reporting the early research; at the middle stage, they just talked about their feelings rather than anything about design or analysis, suggesting that they didn't focus on the key points; finally, they made open reply eloquently and tried to explain the scheme in a situational language, showing their clear logical thinking, and clear reasonable explanation of design point. Compared with the past, they had a qualitative leap.

5 Focusing on fostering the students' practical operative ability

Here it mainly refers to students' ability to express the design effect, namely the ability to evoke the design idea. To gain this ability, students need to receive systematic specialized training, especially the training on software operation and expression by free hand, during their undergraduate studies. The main method is to study independently and practice through long years. Even if a design scheme is very excellent, it amounts to nothing if it is not shown in the writing or the expressive effect is too poor to be incomprehensible. Therefore, the expressive ability of design is another skill essential for a designer.

Free-hand sketch ability helps students to deliberate over their scheme at the early stage. It can directly transfer a designer's thought and design idea, and even express a design emotion, but it is not merely a tool used to transmit the design language, since it has unique artistic appeal. So, like other painting works, sculptures, calligraphy, photography..., it has to meet many artistic requirements. In addition, we regard free-hand sketch as a language exclusively belonging to designers because designers can turn abstract design thinking into something concrete and keep improving it, and free-hand sketch is indispensable from this process, since it can help designers seize the inspiration the most quickly. It may be a common case that sometimes an idea flashes across your mind, and you can capture it at a draught if you are good at drawing. However, if you first switch on the computer and open the software, you will forget it.

The ability to operate software works when the scheme takes shape. It has an advantage that a scene can be restored to the life and gives outsiders the most genuine, intuitive visual perception of your scheme. In particular, the production of an animation and the adoption of VR technology can add luster to the scheme when it is reported. In this way, viewers can not only understand your scheme in the 2D effect picture, but also feel a more real scene in the 3D visual effect.

In general, design teaching should fully meet the needs of the present social development, and give full play to the school's advantages to produce the society new talents that all-round design ability, solid basic skills, a professional spirit and quality, and team cooperation ability.

3 注重校企结合，培养学生对于项目的实际把控和操作

由于我们一组这次选择的是一个实际的PPP模式项目，所以，最终方案是否能够"落地"至关重要，所以，同学们要学会在创意与实际运作之间寻求平衡关系。再加上，该课题本身要求我们必须切实的面向身边的传统民居空间，参与建设，思考其未来发展问题。为该地区的经济和社会发展服务是最终教育教学的宗旨，也为我们提供了最大的"课堂"和施展才华的舞台。

因此，与这次课题密切相关的企业，实现校企结合，是完成人才培养和实施教学任务的一个可靠保证。"室内设计6+1"校企联合毕业设计活动就是一个典型的成功案例，且本届的形式相对于以往来说，效果更加明显。这次的校外合作企业代表，会全程参与和指导同学们的设计工作，在平时的教学中熟悉学生的基本情况，交流沟通无陌生感。在过程中，企业代表会及时纠正同学们一些不切实际的想法和设计内容，指出哪些是好的，哪些是无法实现或产生不了实际价值的，也会把行业中的新材料、新技术、新工艺带入课程教学，为以后学生进入工作岗位，提前打好基础。

4 注重培养学生的逻辑思维能力和表达能力

许多艺术类院校的学生都有一个特点，就是感性思维和发散性思维较强，而逻辑思维及分析能力较弱，我们的学生也有这个特点，这就会造成设计进行时，容易只注重造型和形式感的东西，而不去思考设计的初衷是什么？到底应该解决什么样的问题？缺乏对于设计整体的把控能力。

再有，由于对于逻辑思维能力的欠缺，很多同学在做出方案之后，却不懂得如何将所有设计内容有条理地组织在一起，前后的因果关系混乱，常常会让不了解的人觉得不知所云，这会直接影响项目信息的传递。还有，大部分理性思维较弱的同学，常常也会体现在表达能力不足，这里说的表达能力不光指设计的表达能力，还指语言的表述能力，即与人沟通的能力。设计得再好，但是说不出来，不能让听者完全明白你的意图，那么，这也是一个失败的方案。语言的表达能力是一个设计师必备的技能。

因此，在设计之初，一定要着重培养学生的理性分析和逻辑思维能力，让学生能够通过最初的感性认知逐转化为理性的分析，在不断发现问题和解决问题之间循环往复，多次训练之后，使学生将前期的内容与设计目标直接联系起来，一击即中，直接抓住设计的重点。并且要训练他们能够一步一步将这个设计过程表述出来，即发现问题、提出问题、研究问题、找到解决途径、提出解决方案、通过分析和比对深入设计、深化设计、反向论证设计方案，最终得到最优化的设计方案。在这个过程中，还要不断鼓励他们对自己的方案进行口头论述，不光会做，还要会说，通过不断练习，找到该方案的最佳表述方式，让人一目了然、一听既明。

我们的同学在这次的联合毕业设计活动中感受最为深刻，每一次汇报都能看到明显的进步。由最初调研汇报时的战战兢兢；到中期汇报时的只讲情怀不讲设计过程和分析内容，抓不住重点；到最终答辩时的滔滔不绝，并尝试用情景带入方案讲解，逻辑思维清晰，条理通顺，每个设计点都解释清楚，有理可依。相较以前，可以说是有了质的飞跃。

5 注重培养学生的实际动手能力

这里主要指是学生的设计效果表现能力，即设计想法的实际再现能力。这一能力主要依靠同学们在本科教育阶段系统而专门的训练方法来获取，尤其是软件操作和徒手表现的训练，主要依靠自身的学习和日积月累的练习。一个设计方案想再好，画不出来等于没有；画的表现效果太差，别人看不明白，也等于白做。因此，设计的表现能力也是一个设计师必备的另一项技能。

手绘的能力是帮助同学们在前期推敲方案时使用的，它可以很直接地传递设计者的思想、设计理念，甚至表达一种设计情感，但它又不只是传递设计语言的一种工具，它还有独特的艺术感染力，跟其他绘画作品、雕塑、书法、摄影等一样，它也有很高的艺术要求在里面。再有，我们说手绘是专属于设计师的语言，因为设计师能够把抽象的设计思维，向具体化进行演变、推进，并不断完善，这过程中都少不了手绘，手绘能够最快速的捕捉到我们的设计灵感。不知道大家有没有这种情况，有的时候一个理念就是一闪而过的事情，如果你手绘够强大，就可以很快的捕捉到，并画出来，但如果这时候你去开电脑、开软件，这个信息有可能就忘了，捕捉不到了。

软件操作的能力是帮助同学们在后期方案成型阶段时使用的，它的优点是可以真实的还原场景，令不是行业内的观者通过看你的方案表现，也能得到一种最直观、最真实的视觉感受。尤其是动画的制作以及现在VR技术的加入，可以为设计者的方案汇报加分，让观者不仅能够从二维的效果图中理解你的方案，还可以从三维的视觉效果中体会更加真实的场景效果。

总的来说，设计专业的教学应该紧密结合当前社会发展的需要，结合自身优势，为社会培养设计能力全面和基本功扎实，具有专业精神和素质，以及团队合作能力的新型人才。

A Pilot Study on Organic Renovation of Beijing Qinglong Hutong Community

Yang Lin, Zhu Ningke, Liu Bining, Teng Xuerong
[School of Architecture and Urban Planning, Beijing University of Civil Engineering and Architecture]

Abstract: Returning to its kind with the walls demolished and holes made, Beijing Qinglong Hutong Community (QHC) is calling for further enhancement of its environmental quality. This calls for a study on the feasibility of renovating the QHC from the perspective of localization design, to provide an idea for solving the problem of hutong community innovation faced by Beijing.

Keywords: Beijing, hutong, localization design, renovation

In the only 25 hutong conservation areas throughout Beijing, a great many quadrangle dwellings have been reduced to dead courtyards stuffed with houses, characterizing the adverse living environment by high population density, small per capita living space, and absence of a bathroom and shower in most houses. This is arresting the benign development of the old urban district. A suggestion is put forward from the perspective of localization design in this paper for renovation of the QHC.

1 The Status Quo of the QHC

The QHC, located to the east of the central axis of Beijing, extends east-westward and is about 577m long. It lies to the southwest of Xiaojie Bridge and in the North Second Ring Road, bordered by the Yonghe Temple in the west, at the intersection of Yonghe Tower and the traditional dwelling zone situated to the south of the QHC. Along the street is an old hutong community, and on the other side is a newly-built creative industry office building. In the area, Here are six hutongs parallel to one another in the south and north, which are reconstructed from what used to be a police station. (Fig.1) Among the quadrangle dwellings in this area, some are newly-built pseudo-classic ones, some are well-furnished ones, which are now serving as government offices, and most are "compounds" occupied by many common households. Such a "mixed" dwelling state in this area has representative significance in the whole old city of Beijing.

Fig.1 A map of Beijing Qinglong Hutong Community

1.1 A Research on Qinglong Hutong

A research was launched into the QHC in order to learn about the "mixed" dwelling state in the community. The ways of research included spot survey, resident interview and questionnaire survey. (Fig.2) For the questionnaire survey, 100 questionnaires were issued and collected. Specifically, 88 questionnaires were issued and collected in the compounds (18 compounds in all), while 18 questionnaires were issued and collected in the hutong. 100 households were interviewed.

The questionnaire fell into three parts. The first part was a survey on the basic condition of the resident households, including the date of immigration, house ownership register, family members' age, occupational structure, annual household income, per capita area, basic facilities and living facilities; the second part was a survey on the unruly buildings in the compounds, including whether they had built any unruly buildings, for what purpose they used these buildings and whether they supported in demolishing these buildings; the third part was a survey on the

residents' willingness, including whether they were satisfied with their living condition, as well as the cause of dissatisfaction, whether they were willing to spend money to renovate the house, and whether they were willing to participate in the planning and design.

1.2 The Extant Problems of the QHC

As shown by the research, the living space was generally cramped in the QHC, and most households' living standard was low. When a house went wrong, it could only be repaired temporarily, while the problem couldn't be solved completely. In the compounds there was no underground cable, but wire poles erecting on the roads, and wires intertwined with one another, which passed through the houses and might cause trouble in safety. The pivot of some gates had got rusty, making the gates hard to close, thus reducing them to something of no use; the bricks in some walls had fallen off. Most of the residents were old people, who didn't have any recreational facilities or place to chat, take a walk or do exercise. Neither was there any space for children to play in (Fig.2).

2 Localization Design Principles

The word "localization" has something to do with "a certain local region", especially with the local regional characteristics. "Local culture" is, as it were, a national common view formed by the culture and history of a certain region in the unique environment of this region. In mankind's memory, what's unforgettable is not a "scene", but a "place" full of feelings. For a culture containing only internal cultural factors rather than both internal cultural factors and collision factors, it can never develop into a pluralistic culture. The root in which a local culture is rooted is closely related to the local residents and their most practical "community life". As the best embodiment of people's insights on the plainest life and inheritance of the "local

北京青龙胡同街区有机更新初探

杨琳 朱宁克 刘璧凝 滕学荣

摘要：北京青龙胡同社区经过整治拆墙开洞恢复胡同风貌后，需要进一步提升环境质量。以在地设计观点探讨北京青龙胡同社区更新的可行性，为解决北京胡同社区改造问题提供思路。

关键词：北京 胡同 在地设计 更新

图2 青龙胡同街区实景

北京老城仅存的25片胡同保护区中大量四合院已经沦为盖满了房子的死院，人口密度高、人均居住面积小、大部分家庭没有卫生间和淋浴设施，人居环境恶略，抑制了老城区的良性发展。本文以北京青龙胡同为例，以在地设计观点，提出北京青龙胡同社区更新建议。

1 北京青龙胡同街区现状

北京青龙胡同街区位于北京中轴线东侧。青龙胡同东西贯穿，长约577m。位于北二环内小街桥西南侧，西临雍和宫，处于歌华大厦、雍和大厦和青龙胡同南边的传统民居的交汇区域。沿街一面是北京现有的胡同老街区，而另一面则为新建的创意产业办公大楼。该区域原为炮局拆迁改建而成南北各有6条胡同排列（图1）。该地段内现有的四合院中，既有新建的仿古四合院，也有条件较好的四合院作为各机关单位办事机构，但数量最多的是普通居民住户多户杂居的"大杂院"。该地区内的"大杂院"的生活状态，在整个北京旧城中具有代表性的意义。

1.1 北京青龙胡同调研

为了解北京青龙胡同街区"大杂院"中的居民的状况，对北京青龙胡同社区进行了调研。调研方式包括实地踏勘、居民访谈和问卷调查（图2）。其中问卷调查共发放收回问卷100份。其中院落内发放收回88份（共18个院落），胡同内发放收回18份。受访家庭共计100户。

调查问卷分为三个部分。第一部分是对居民住户的基本状况的调查，包括迁入时间、房屋产权、家庭人口年龄、职业构成、家庭年收入、人均面积、基础设施、生活设施八个方面的问题；第二部分是对院内加建的调查，包括是否加建、加建面积、加建用途、是否支持拆除加建四个方面的问题；第三部分是对居民意愿的调查，包括对居住情况是否满意、不满意原因、是否愿意投资修房、是否愿意参与规划设计四个方面的问题。

1.2 北京青龙胡同街区现存问题

调研显示，大部分青龙胡同街区内住户的住房面积局促，居住房间面积都很小，住户生活水平普遍较低。当房屋出现问题时，只能是临时的修补，无法从根本上解决问题。没有电缆入地，电线杆竖立在路上，错乱的线路交织在一起，接入居民房屋内，存在安全隐患。一些院落大门的门轴已经生锈，很难关上，只能作为摆设；有的墙壁上，已经有砖块脱落。此处居民老龄化程度较大，老年人没有文化设施休闲、聚集聊天和散步健身的活动场所。小孩子也没有可以活动的场所（图2）。

2 在地设计原则

在地二字，"地"是指"地域"，根植本土地域特色的形态可称在地。"在地文化"可以说是某一地域上独有的环境，人文、历史等一切面向在此区域中产生的民族性共识。在人类的记忆中，令人难以忘怀的不是"景"，而是"境"，及充斥着感情的场域。只有人文内置因素而没有碰撞刺激的文化不足以发展成多元性。在地文化的根植土壤是与其地域的"人"及最真切的"群体生活"息息相关。透过地域文化认同，导引"以人为本"及"联结生活"的体验，借由

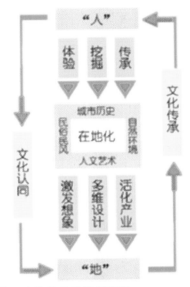

图3 在地化设计发展流程

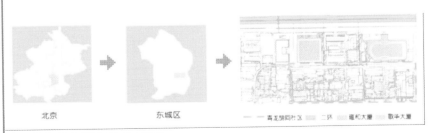

图1 北京青龙胡同社区区位图

culture", local culture identification can help to experience the "community life" with people at the core, and dig deep into the valuable local culture of a city, including its history and folk customs, with people as a medium. This may arouse an intuition and passion for the local culture and art, to promote cultural inheritance, so as to bring new inspirations and opportunities for city innovations (Fig.3).

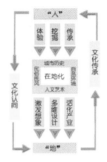   

Fig.3 the development process of localization design

Fig.4 A real picture of Zhengyang Bookstore

Fig.5 A Picture of the old Beijing folklore collected in Zhengyang Bookstore

Fig.6 A public lecture on the culture of Beijing held in Zhengyang Bookstore

3 An Exploration on Organic Renovation of the QHC

3.1 Integration of the Old Beijing Culture into the Hutong Community Life

When applied to design, the word localization has very extensive implications. Based on it, as well as the profound culture of Beijing, innovation is activated, from which cultural elements are extracted and properly translated into symbols and characteristics and integrated into creativity, in a bid to convey the unique connotations and value of the Beijing culture during renovation.

Zhengyang Bookstore is located in Zhuanta Hutong, or rather, in the courtyard where Pagoda of Monk Wansong is (Fig.4). Zhuanta Hutong was founded in the Yuan Dynasty, and there is literature evidence for this. As the only case in Beijing, it is the "root" of the hutongs in Beijing. Zhengyang Bookstore is bent on protecting, developing and utilizing the historical documents of Beijing as a supplement to the institutions and historical data, in a bid to preserve the documents of the old Beijing and inherit the culture of Beijing. "I found the publishing house just because I like the culture of Beijing and want to inherit it," said Mr. Cui, a native of Beijing, six generations of whose family have lived in Beijing. He admits that he has a deep feeling for Beijing and its culture. Most of the books in Zhengyang Bookstore record the historical evolution of Beijing objectively. The frequent visitors of the bookstore are native citizens of Beijing and the readers interested in the history of the old Beijing. To inherit the culture of Beijing in more ways and show more people the culture of Beijing, Cui Yongsi, the owner of the bookstore, keeps thinking up new ideas and trying new carriers. Zhengyang Bookstore holds various book fairs, photo exhibitions and public lectures about the culture and memory of Beijing, to communicate the culture of the old Beijing by word of mouth.

Zhengyang Bookstore is not only an ordinary bookstore, but a very convenient platform that offers an understanding of the culture of Beijing and makes more people willing to step into Zhuanta Hutong to learn about the old Beijing; it has developed an inseparable relationship with the people in the hutong and in the whole Beijing city. It relates people to a region through cultural identification and guides them to dig deep into the history of Beijing city, so as to focus on the inheritance of culture; it is a typical case of localization design of hutongs.

3.2 Preservation of Regional Culture Characteristics in the Renovated Hutong Block

Not only is localization design a practical application of the essence of design service, but it helps to re-experience the beauty of a region. Also, design media and pluralistic cultural integration make design closer to everybody who has a relation with the region. And with renovation guided by the regional culture, the region's uniqueness can be better revealed.

As one of the oldest hutongs in Beijing, Shijia Hutong was founded in the Yuan Dynasty and officially named in the Ming Dynasty. The courtyard pattern of the only hutong museum here was still completely preserved. The courtyard of the museum was then renovated in order to maximize the restoration of the archaic charm of Shijia Hutong. Most of the tiles, including floor tiles, were collected from the native households in the hutong, and the renovation was performed with the courtyard unchanged. The renovated museum has a double courtyard, covers an area of over 1,000m2, and is composed of 8 exhibition halls and 1 multi-function hall. (Fig.6) The historical features of the hutong are displayed, and not only so, but favorable conditions will be provided in future for the residents in the hutong to carry out activities. Too many memories and sounds of the old Beijing are preserved in the hutong museum. Elementary school students often pay a visit to the museum, where the history and hutong culture of the old Beijing are propagandized (Fig.7).

Intangible cultural heritage inheritors also tell about the peddling sounds that could be heard in the streets and lanes of Beijing last century, to vividly show children a live history that contains the authentic charm of Beijing. It is a living museum, which exists in order to record history, to make the contemporaries and later generations more willing to stay close to history. In this case, a hutong community was renovated based on the local culture, and that's why the design is called localization design.

3.3 Pluralistic Integrating Innovation as an Inspiration for Renovation of Hutong Communities

Interdisciplinary design collaboration can help to integrate design into people's life from multiple perspectives to inspire the imagination, so as to comprehend the region in a new way, practice and activate the local industries and circulate it in design

人为媒介深入挖掘城市历史，民俗民风等极具人文价值的宝贵区域文化，从而激发对在地人文艺术的直觉和热情，转而为文化的传承，为城市的创新带来新的灵感和契机，这是人最平实的生活的体悟与"在地文化"承袭的最佳体现（图3）。

在地文化的多元化迫切需要跨领域的设计协作，多视角地将设计融入人的生活之中，激发想象力，继而对场地有新的体悟和理解，实践活化地方产业并循环于设计创新，从而达到设计永续化。站在保留在地文化的角度上，把自然、社会、科技、艺术等融于地域的多维度设计，就是"在地设计"。

3 北京青龙胡同街区有机更新探索

3.1 老北京文化融入胡同街区生活

在地化之定义于设计，内含极为广泛，以此为源泉，依托为底蕴深厚的北京文化，活化创新业态，从中摘取文化元素并适当转化为符号和特征融入创意之中，在更新的同时传达北京文化的独特内涵和价值。

正阳书局坐落于砖塔胡同"万松老人塔"所在的院落（图4）。砖塔胡同是从元代到今天都有文献可考，这在北京是孤例，它是北京胡同的"根"。正阳书局的初心是致力于北京历史文献的保护开发及利用，补典章之得失、史乘之缺憾、保存老北京文献、传承北京文化。"我办这书局原因很简单，就是因为喜欢北京文化，想要传承这文化。"崔掌柜是个地地道道的老北京，家中六代人都在北京生活，他坦言对于这座城市、对于这里的文化他有着很深的感情。正阳书局的大部分藏书是从历史上真实、客观地记录了北京的发展变迁。书店的常客是土生土长的老北京人和被老北京的历史感兴趣的读者。为利用更多的方式传承北京文化，让更多的人了解北京文化。书店主人崔勇思考新办法、新载体。正阳书店现在也举办关于北京文化、北京记忆的各种书展、影展和公益讲座，用口头记述的方式传播老北京文化（图5、图6）。

正阳书局的存在，不仅仅是一个平常的书店，它已成为了解北京文化非常便利的一个平台，也让更多的人愿意走进砖塔胡同了解老北京；它与这个胡同里的人，整个北京城里的人发生了密不可分的联系，它通过文化认同引导人与场所的联系，深层挖掘北京城的历史从而落脚于文化的传承；它是胡同在地设计更新的典型案例。

3.2 胡同街区更新保有场所地域文化特色

在地化设计不仅仅是对设计服务本质的务实致用，也是透过设计重新体验地域的美感，以设计媒介和多元整合文化，从而使得设计更加贴近每一个与这个场所有关联的人。根据场所的地域人文引导更新方向，可以更好地显示场所的差异性和独特性。

作为北京最古老的胡同之一，史家胡同始建于元代，明代正式定名。在此处创立的惟一一家胡同博物馆院落格局保存十分完整。为最大限度地还原了史家胡同古韵，博物馆院落翻新中，包括地砖在内的大部分砖瓦都是从胡同老住户中征集而来，在不改变院落格局的前提下进行翻新。翻新后的博物馆为两进院落，占地面积一千余平方米，设有八个展厅和一个多功能厅（图7）。除展示胡同历史风貌外，今后还将为胡同居民开展活动提供便利条件。胡同博物馆保存着太多属于老北京的记忆和声音。博物馆经常开展小学生参观活动，旨在向下一代传达老北京的历史和宣扬胡同文化（图8）。并有非物质文化遗产传承人口相传上个世纪北京街头巷尾的叫卖号子，用活的历史使孩子们更加深刻的体会到原汁原味的北京风情（图9）。这是一座活着的博物馆，博物馆的存在正是为了记录历史，让当代的人以及我们的后代更愿意亲近历史。这个案例充分以地域人文为导向进行了胡同社区的更新，只有这样，才能成为真正的在地设计。

3.3 多元融合创新为胡同街区更新创造灵感

跨领域的设计协作，多视角的把设计融入人的生活之中，激发想象力，继而对场地有新的体悟和理解，实践活化地方产业并循环于设计创新，将各个表现形式有机地编织到胡同特殊的城市肌理及有趣的院落空间之中，通过空间改造、社区生态建设及环境提升、本地区手工艺探索及各门类艺术设计，持续加

图 4　正阳书局实景

图 5　正阳书局旧京风俗展

图 6　正阳书局京味文化公益讲座

图 7　胡同博物馆实景　　图 8　小学生参观胡同博物馆　　图 9　手艺传承人

innovations, and fuse various forms of expression organically into the special urban texture and interesting courtyard space of the hutong. Then, space transformation, community ecology construction and environmental upgrading, local handicraft exploration and art design will continue to deepen the dialogue on the protective renovation and transformation of the QHC.

The Dashilan and East Liulichang Historic Conservation Area, located to the southwest of Tiananmen, is the historic conservation area nearest to Tiananmen and one of the largest ones where there is the widest variety of historical remains that are best preserved. Beijing Dashilan is a main target of zone protection and revitalization, for which the urban planners, historical and cultural protection experts, architects, artists, designers and commercialists cooperated with one another to explore a new way to protect the Dashilan Historic Conservation Area.

The Dashilan renovation program was launched in 2011, and then a journey to design was held during the Beijing International Design Week. Excellent Chinese and foreign design and artistic creative projects were invited to the old street block, providing a new idea for the renovation and activation of historic conservation areas—exploring new use of old houses on the premise of respecting the texture of old street blocks, and introducing new industry patterns on the strength of design to create new communities. At the Journey to Design 2015, the "Dashilan Community Design" project will be launched in order to: meet the community demands and promote the community development; provide a series of community services and activities to facilitate the planning and development of the public space, to enhance the community cohesion and regional sense of pride; (Fig.10, 11) many participants, organizations, institutions, local residents and dealers will participate in the "Journey to Design—Dashilan Design Community" 2015 to continue deepening the dialogue on the protective renovation and transformation of Dashilan. Some groups will become the new owner of the Dashilan Community and reside in the street block forever, to promote its prosperity and revitalization.

Fig.10 Beijing Design Week—Dashilan Dubai Design Show 2015   Fig.11 Beijing Design Week—Dashilan Peripherals 2015

4 Conclusions

This paper mainly analyzed the problems currently faced by the hutong communities in Beijing from the perspective of localization design, and then gave advice on the renovation of the hutong communities in Beijing. Localization design can better retain the distinctiveness of a regional culture and the continuity of the historical context. Pluralistic integration and collaboration, regional cultural activation and innovation can help to promote the practice of localization design. Beijing has a long history and culture. Under the principle of localization design, the identifiability of the hutong communities in Beijing will be better highlighted and they will be further distinguished; so, only localization design can provide a sustainable road for the renovation and transformation of the hutong communities in Beijing.

References

[1] Du Ruize. Design according to Circumstances: A Rustic Opinion of "Localization Design"[J]. Art Observation,2015(1): 16-17.

深青龙胡同社区保护更新与改造的对话。

大栅栏及东琉璃厂历史文化保护区（大栅栏及东琉璃厂文保区）位于天安门西南侧，是离天安门最近的、遗存遗迹最丰富、保留最完好、最大的历史文化街区之一。北京大栅栏作为区域保护与复兴的实施主体，通过与城市规划师、文化历史保护专家、建筑师、艺术家、设计师以及商业家合作，探索实践大栅栏历史文化街区的保护新模式。

自从2011年发起大栅栏更新计划，与北京国际设计周举办设计之旅，邀请中外优秀的设计和艺术创意项目进驻老街区，为历史文化街区的更新活化提供了新思路——在尊重老街区肌理的前提下，探索老房子新利用，通过设计的力量引入新业态、营造新社区。2015年的设计之旅"大栅栏设计社区"的项目主要围绕以下三个板块：通过设计满足社区需求与促进社区发展的产品整合；提供一系列社区服务与活动，帮助公共空间的规划与发展，促进社区凝聚力及区域形象自豪感；设计与环境的共生，进行社区建筑、公共环境与设施的改进和完善（图10、图11）。2015年的"设计之旅——大栅栏设计社区"将有多个参与者、团体、机构、在地居民及商家共同参与，持续加深大栅栏保护更新与改造的对话，部分群体也将成为大栅栏社区新主人永久入驻街区，成为繁荣复兴的种子，实现活化老街区的新景象。

4 结论

本文主要从在地设计的角度分析了北京胡同街区的现状问题，进而引导北京胡同街区的更新。在地设计可以更好地保留地域文化的特殊性和历史文脉的流传性。多元融合协作、活化地域文化、创新整合则有助于促进在地设计的实践。北京具有悠久的历史和文化。在地设计的原则下，会更好地彰显北京胡同街区的可识别性并提升差异化，为北京胡同街区的更新改造提供一条可持续发展的道路。

图10 2015北京设计周大栅栏迪拜设计展　　图11 2015北京设计周大栅栏周边产品

参考文献

[1] 杜瑞泽. 因地而设计："在地化设计"刍议[J]. 美术观察，2015（1）：16-17.

Practice and Reflection on Spatial Architecture Topics

Lu Qinzhi, Song Yang
(School of Arts, Zhejiang University of Technology, Hangzhou 310032, Zhejiang)

Abstract: Basic design curriculum makes sense since it can help students to integrate the art foundations and advanced design courses together, offer them a progressive understanding of how to learn design step by step by starting with fundamental design methods and rules, then give them a comprehensive understanding of the design methodology and design-related knowledge, and enable them to gradually improve and build a design thinking system of their own. For the students majoring in space design, model making is not only a skill, but also a way of pondering problems. The course begins with model making to gradually adapt students to the basic conceptive pattern "conceiving—sketching—preliminary model making—model making" when doing coursework, and gradually teach basic course learners, who are just accustomed to "imagery thinking", to replace plane thinking with stereoscopic thinking and formative thinking with spatial thinking". The research of space architecture topics makes sense because the combination of thinking methods with hands-on operations can enable students to gain insights in practice, get feelings in the process, and promote the development of practice with their acquired design thoughts.

Keywords: Space design, classroom teaching, design basis

Design differs from art, at least at the level of basic design curriculum. So, teachers should make a clear distinction between design and art, or students won't define the ultimate goal of design as "identifying and solving problems", let alone emphasize a "design method" as an "approach" developed from a constantly improved idea gained from constant "restrictions" and practice and discriminate it from pure "creativity". Over ten years of experience in teaching basic design courses make me deeply aware of the difference between design and art in design basis (in teaching idea and method). The difference in itself lies in the total difference in aim between design and art. The significance of design lies in the way of solving problems and the result finally presented, while art aims to verify with ideas and methods time and time again that there is no standard or rule for art. The teaching aim of spatial architecture curricula is to familiarize students with the basic idea and method of model making (linked with a systematic conceptive pattern); to learn to think in the manner of "model", and gradually improve the stereoscopic and spatial thinking model, to finally build a perfect spatial thinking system.

1 The teaching content is constantly improved and renewed in order to meet the requirements of design

I have been improving the spatial architecture topics for the following purposes: first, to make a knowledge supplement and improvement in order to link the basic design courses with specialized design courses; second, to make the "spatial perception and spatial organization" in the basic design curriculum, which is very important to space formation modes, into the connection point between generalized design foundation (general education foundation and fine art foundation) knowledge and specialty foundations (basic design courses), and develop it into a perfect curriculum system. I will also make an overall topic design and practice to verify my teaching idea, so that the students, who are now studying the basic design courses, would learn the specialized courses easily.

Practice of 7 teaching topics:

Topic 01: space representation
Sketch a large stack of white geometries with a pencil by combining light with shade; then begin to sketch a single-colored architectural space (different drawing materials and expression techniques can be adopted).

Topic 02: an equally divided cube
Adopt the equal dividing method as the most direct means of space acquisition to generate 2 interrelated spaces, to provoke the students' thinking about the problems related to spatial modulus and space division.

Topic 03: a block box
Pay attention to the process in which 2D graphics develops into a 3D "spatial shape".
Divide a cube with a given volume into several pieces, and then reassemble the pieces together to build a new spatial entity.

Topic 04: a storage box
Create a closed space that can be opened and hold articles, to provoke the students' thinking about the interaction between the "virtuality and reality" of the space, between the "positivity and negativity" of the spatial shape, and between the shape and the articles.

Topic 05: hand and box
With the intervention of space creation and scale, mobile phone can serve as a benchmark for space scale or part of spatial structure.

Topic 06: Kandinsky's space
Topic 07: tandem boxes
Divide the inner space of a cube with a given volume into 3 tandem spaces.

2 A reflection on the problems related to course teaching

2.1 Space design is a rational game

Space design must be perceptual, because people will enter the space to feel what is conveyed by the designer with their eyes and body, while they won't look at the designer's conception in words; space design must be rational, because design makes sense only when it's used to solve problems, and many space problems arise from the perception of these problems,

空间构筑系列课题的实践与思考

吕勤智　宋扬

摘要：设计基础课程的意义在于使学生对前置艺术基础可以与后置设计课程的对接，循序渐进的使学生通过对设计每个环节的基本设计方法、设计规则的了解切入设计学科的学习，并逐渐对设计方法论、设计相关学科知识有全面的了解，并逐步完善、构建自身的设计思维体系。制作模型对于学空间相关设计专业的学生来说，既是一门技能，又是一种思考问题的方式。课程以模型制作为切入点，使学生从课程作业的过程中逐步适应"构思→草图→草模→模型"的基本构思方式，从基础课学生只是习惯运用"形象思维"这一种思维方式，逐步转换为"从平面到三维、从造型到空间"的思维方式。空间构筑系列课题研究的意义在于用思维方法与动手实操的互相结合，使学生在实践中领悟，在过程中感受，用获取的设计思路推动实践的发展。

关键词：空间设计　课堂教学　设计基础

设计与艺术不同，至少在设计基础课程的层面上，作为教师应该把设计与艺术的关系明确分开，否则学生不会把设计课题的最终目的定义为"发现问题与解决问题"，更不会强调"设计的方法"是在不断的"制约"中，经历实践获得的思路，并不断完善而发展出的"方法"，与单纯的"创意"不同。十余年设计基础课程教学经历，使我深刻的体会到设计与艺术在设计基础课题层面的差异性（在教学思路、教学方法上的差异性），差异性的根本在于设计与艺术的目的完全不同，设计的意义在于解决问题的方式与最终呈现的结果，而艺术的目的在于用思路与方法不断的一次次验证艺术是没有标准、没有规则的。空间构筑系列课程的教学目的，首先是让学生熟悉模型制作的基本思路及方法（与系统化的构思方式衔接）；其次是学会用"模型"的方式思考，并逐步完善立体与空间的思维模式，最终构建完善的空间思维体系。

1 教学课题的不断完善与更新，是为了对接设计课题的需要

空间构建方式的系列课题是我一直在不断的完善与改进的，改进的目的：一方面为了衔接设计基础课程与专业设计课程的需要；另一方面，我设想把设计基础课程的"空间感知与空间构建"这个对空间生成方式至关重要的环节，称为广义设计基础（通识基础、美术基础课）、知识与专业基础（设计基础课）衔接阶段的连接点，并发展成完善的课程体系。进行全面的课题设计与实践对教学思路进行验证，使学生从设计基础环节向专业学习可以顺利地过渡。

7项系列教学课题实践内容如下。

课题1：空间的表述
用铅笔素描的方式，以光影表现与体量表现相结合的方式，借助大体量白色几何体的堆砌作为写生对象；并逐步过渡到单色（可选择不同绘画材料、不同表现技法）的建筑空间写生。

课题2：等量分割的正方体
用等量分割作为最直接的空间获取方式，生成两个互相关联的空间，引发学生对空间模数、空间分割相关问题的思考。

课题3：积木盒子
以既定体积的正方体为原形，分割正方体得到单体，并重新寻找组合方式，构建新的空间造型。

课题4：储物盒子
创作可以开启，装载物品之后，闭合的空间，引发学生对于空间的"实与虚"、空间造型的"正与负"，以及空间与装载物品之间的互动相关问题的思考。

课题5：手与盒子
空间创造与比例的介入，手机可以作为空间比例的基准，又可以成为空间构造的一部分。

课题6：康定斯基的空间
关注从二维"图形"向三维"空间造型"发展的过程。

课题7：串联的盒子
限定体积的正方体为原形，进行内部的空间划分，把空间分割为互相串联的三个空间。

2 课程教学相关问题的思考

2.1 空间设计是理性的游戏

空间设计必须是感性的，因为人需要置身其中，用视觉、身体去感受设计

while the result of design is nothing more than a way of solving these space problems, and the space's function can serve to verify the designer's design idea. So, space design needs a creative idea, but every creative idea must be justifiable.

The space language can be perceptual, but a space architecture idea must be a thought derived in various restrictive conditions, and the solution to many problems may be the carrier of the creative idea, so after a perceptual creative idea is permeated into the many details of space thinking development, the final space can provide a deeper experience feeling in detail.

2.2 "Rule learning makes sense when the rule is broken"

At the level of "design basis", topic practice does not permit students to "match" a coursework topic with the final work untimely or match the creative idea with a certain design type untimely, or look for a space creation idea or method in a single design type or thinking pattern. Students should be prevented from having their creativity tied to some specialty thinking, and their way of designing and solving problems confined to a single pattern. On the contrary, a unique idea may be able to highlight a design work.

"Every creative idea means an adventure, which may redefine the design work." Despite the importance of the basic design courses, it may be more important to define a method and standard as design deepens. However, facts prove that a result of thinking based on a given rule in a given way is mostly "expected". In basic design teaching, rule learning has almost spilled over into all the courses, such as composition, drawing, surveying and mapping, and masterpiece analysis. These regular required courses have enabled me, as well as many other students, to work in a company in the society, but rule learning has nipped students' creativity "in the bud". A validated teaching idea, which is developed from teaching processes, is mostly a "rule-based one, however it won't make sense unless the rule is broken".

Almost everybody experiences something alike in his learning process. When studying at university, one usually has eyes only for the works in which modeling techniques are applied in the best way, such as Tschumi's work, in which the rule of form is redefined in the deconstructionist way, Eric Owen Moss's work, in which beauty of section planes tells the world that only the structural collision is the essence of the architectural form, Peter Eisenman's work, in which the modernistic definition of architecture is negated by the modernistic formal rule, and Fujimoto's work, in which the author reflects upon the essence of architecture on the fringe of shape and function, and even brings architecture back to the starting point of all beauty—a beautiful shape comes from nature. The late architect Zaha verified the "boundlessness" of spatial space with product design, costume design and industrial design. The boundary in itself refers to the scale.

2.3 The best way to avoid misreading is "draw no conclusion"

"Design in Japan features the artistic conception of Zen...the timber structure is fired with the oriental spirit..." Our mind is inundated with too many "fantastic" definitions. If we cannot give an accurate definition, we would better not anxious to define it. This is because many design ideas and methods are under the absolute individual category, and there is no standard or answer, so it is better to find "something useful" in others' works than to get a wrong answer. So teaching makes me deeply aware of the importance of "not defining any work used for reference". The students hold in their hands a lot of ideas and data obtained from the Internet, which need to be understood slowly by the students with the help of the course teacher. Most often, a teacher's answer to a student's question, which is usually a definition of reference works, has a great impact on the development of the student's idea. However, a reflection on the talks with many designers from Japan, Europe and America shows that their works are defined erroneously from time to time.

My mentor Bahman, a winner of the European Lifetime Achievement Award for Architectural Art, said to me, "When I was young, I liked the architectural shape; when I grow older, I try to simplify the shape, and think nothing is more important than living in a comfortable residence." I saw many buildings designed by my mentor in his early days. In detail, there are many traditional European-style ornaments in the works, but his residence is just a square building. In the past, many architectural theories and architectural styles might be thought of, but in my mentor's eyes, a "comfortable residence" was the best thing. In the classroom teaching, I just give the students detailed guidance on the formal language in reference works rather than define any work according to the design idea and spirit hidden in it, so as not to hinder them from expanding their thoughts.

3 Conclusions

All teaching practice in classroom must come from the teacher's experience gained in design practice, from his understanding of the subject teaching system, and from the constant revision of the teaching topics made based on student feedback; with the social development, design is redefined again and again. Accordingly, the teaching topics need to be renewed, and teachers' teaching idea needs to be updated as well, so as to meet the design standard constantly updated with the social development. The practice and research of the topics on Spatial Architecture is an exploration on the teaching objective and method of classroom teaching topics at many fundamental design stages. The research on topics and the reflection on topic practice are never ceased, but gradually enriched. To help students "ponder space problems", a complete teaching topic system should be established at the teaching stage, to build them a rounded spatial thinking system.

References

(1) Kenya Hara. Education on Desire[M].translated by Zhang Yu. Guangxi Normal University Press, 2012.
(2) Kenya Hara. White[M] .translated by Ji Jianghong. Guilin:Guangxi Normal University Press, 2012.
(3) Tanaka Ikko. Awakening of Design[M].translated by Zhu E. Guilin:Guangxi Normal University Press, 2009.
(4) Jia Beisi. Architectonics of Modernism[M]. Beijing:China Architecture & Building Press, 2003.

师所要传递的感受,而不是设计师把构思用语言与文字传递给使用者;空间设计必须是理性的,因为设计的意义在于解决问题,空间众多问题的出发点始于对空间问题的感受,而设计的结果止于对空间问题的解决方式,空间需要用功能以验证设计者设计思维的存在。所以,空间设计需要创意的融入,但是,每一个创意都需要理由。

空间语言可以是感性的,但是空间构筑思路必须是基于各种条件限定后的思维推导,众多问题的解决方式往往是创意的载体,把感性的创意渗透在空间思维发展的众多细节中,可以使最终空间的呈现具有更多可以深入体验的细节。

2.2 "对规则的学习,意义在于打破规则"

从"设计基础"层面思考,课题实践中不能让学生过早的把课程作业课题与最终作品完成后的结果"对号入座",过早的把创作的思路对应某种设计类别,从局限于某种设计类别片面、单一的思维模式中去找空间创作的思路与方法,避免过早的使学生创造力被所接触专业的某种既定思路所束缚,设计的方法、解决问题的思路并没有唯一的模式,恰恰相反,与众不同的、另辟蹊径的思路可以成为设计作品的闪光点。

"每一次创意的融入,都是一次冒险,都是设计作品被重新定义的可能性。"设计基础课程的规则固然重要,在往后的设计道路上往往会成为一种方法和标准的界定,但是事实往往却证明,按照既定思路、既定规则思考的结果,往往会在"意料之中"。设计基础教学中,对规则的学习几乎渗透了各个课程,构成、制图、测绘、大师作品分析,这些长久以来的必修课程,把包括我在内的众多学生从学校送到了社会、送到了公司,但是,规则的学习,往往把学生的创造力"抹煞在摇篮中"。总结很多教学的过程,往往事实验证的教学思路是"对规则的学习,意义在于打破规则"。

每个人经历的学习过程中,经历过几乎雷同的阶段,学生时期对造型发挥很极致的作品往往情有独钟——屈米的建筑用解构的方式重新定义了形式的规则,埃瑞克·欧文莫斯的建筑用剖切面的美感告诉世人结构的碰撞才是建筑形式的本质,彼德·艾森曼用现代主义的形式法则打破了现代主义对建筑的定义,藤本壮介在造型与功能的边缘思考建筑的本质,甚至又使建筑回到了所有审美的起点——美的造型源于自然界;回想起已故建筑师扎哈用她的产品设计、服装设计、工业设计,验证了空间造型的"无界限",所谓的界限,仅仅是比例问题。

2.3 避免"误读",最好的方式是"不要结论"

"日本的设计充满禅意……木结构的运用具有东方精神……"我们的意识里充斥着太多我们自己"幻想"的定义。如果我们无法准确定义,最好的方法就是不要急于定义。因为很多设计的思路、方法都是绝对个人化的,没有标准、没有答案,相对于在别人的作品中找到"自己需要的东西",比得到一个错误的答案要好很多。所以在教学中我深刻体会到"不给借鉴的作品下定义"的重要性。学生有很多来自于网络的思路与资料,都需要任课教师用自己的理解帮助学生慢慢体会,很多时候教师对学生的相关提问,所给予的针对借鉴作品定义会对学生思路的发展产生很大的影响,但是反思与众多日本、欧洲、美国设计师交流的时候,我发觉自己很多对他们作品"定义"都是错误的。

我的导师,欧洲建筑艺术终身成就奖获得者巴赫曼·祖尔坦对我说:"年轻的时候,喜欢建筑的造型,等年纪大了,会把造型做得越来越简单,会觉得住着舒服才是最重要的。"我看过导师设计的众多建筑,年轻时候的设计,细节也会有众多传统欧式的装饰,但是自己住宅的设计却只是一个方形的建筑外观,这在以前,也许会联想到众多建筑的理论、建筑风格的发展,但是在导师眼中,仅仅是给自己一个"舒适的居所"。在我的课堂教学中,我只针对借鉴作品的形式语言给予学生细节指导,绝不对作品所反映的思路、所呈现的设计思想及精神层面给予"定义",避免无意中对学生思维的拓展产生桎梏。

3 结论

众多课堂教学实践的必须源自教师在设计实践中获得的经验,必须源自教师对学科教学体系的理解与思考,必须源自学生教学反馈对教学课题的不断发展与修订;社会的发展对设计不断提出新的定义,教学课题需要发展,教师的教学思路同样需要更新,以应对社会发展对未来设计领域供作者不断更新的标准。《空间构筑》系列课题的实践与研究,就是众多设计基础环节,对课堂教学课题的教学目的、教学方法的探索。课题的研究、对课题实践的思考一直在持续,并逐步完整、充实,针对学生"空间问题该如何思考"教学环节的形成完整的教学课题体系,为学生构建完整的空间思维体系。

参考文献
[1] 原研哉. 欲望的教育[M]. 张钰,译. 桂林:广西师范大学出版社,2012.
[2] 原研哉. 白[M]. 纪江红,译. 桂林:广西师范大学出版社,2012.
[3] 田中一光. 设计的觉醒[M]. 朱锷,译. 桂林:广西师范大学出版社,2009.
[4] 贾贝思. 型和现代主义[M]. 北京:中国建筑工业出版社,2003.

專家講壇
Introduction of the Activity

CIID『室内设计 6+1』2017（第五届）
校企联合毕业设计
CIID "Interior Design 6+1" 2017 (Fifth Session)
University-Enterprise Cooperative Graduation Design

国匠承启卷
——传统民居保护性利用设计
Craftsmanship Heritage
——Design for the Protective Utilization of Traditional Folk Houses

國匠承啓卷——傳統民居保護性利用設計

The concept of "urban renewal" originally spread from the West to China in the 1980s. At first, "urban renewal" focused on "metabolism" and "organic renewal" which call for the protection and development of the urban physical environment. After the 21st century, the focus shifted to "urban regeneration" and "urban revitalization" which view urban renewal from diverse perspectives such as culture, economy, society and politics. The renewal of cities with old blocks, which involves more stakeholders, heavily depends on the stakeholders' sense of value, and the balance of interest. According to the urban renewal experienced by Guangzhou, this speech will analyze the decision-making mechanisms developed for various specific stages, and discuss the rights, obligations, and functions of involved stakeholders (e.g. the government, the market, and residents) based on related cases, for the purpose of finding out the future direction of urban renewal in a practical way.

The decline of old blocks can be explained from economic and social points of view. Economically, many urban blocks failed to return to their built environments after releasing regional value, which led to considerable outflows of interest. Consequently, the built environments decayed due to the lack of capital injection, and were finally replaced physically in an intensive way, or continued to deteriorate physically by reason of their low quality. Socially, physical replacement caused gentrification, commercialization, and also fracture of social structure, and physical deterioration led to poverty and social decadence. In either way, old built environments, even those appeared in 1980s-90s which are not self-sustaining, have been existing for some reason.

One thing that is for sure is that the government has been playing an increasingly important role in Guangzhou's renewal, with core decision-making mechanism changing from market-driving renewal to government-decided renewal, and to government-led renewal. Though coming with moderate correction and exploration, this process finally led to rational coordination among related parties. It is worth mentioning that in the process of urban renewal, related parties have been involved in decision-making, and more objectives covering various aspects have been gradually added by the government. In 2016, Guangzhou Urban Renewal Bureau was established based on the original Three-Old Reconstruction Office as China's first Level I urban renewal authority, for the purpose of realizing the urban revitalization of Guangzhou through micro-reconstruction and micro-renewal which are more flexible for urban and social construction. In addition, sophisticated funds, responsibility, authority, decision-making and execution systems have been established to support urban renewal. At present, an obvious problem affecting urban renewal is the public's lack of participation and focusing too much on material interests. Therefore, more experience is needed to solve the problem.

In the current social context, urban renewal should focus on improving built-environments while protecting stakeholders' interests in a fair way widely recognized across the society, and, as a part of sustainable social development, promote the self-renewal of communities. Urban renewal, which comes with rapid urbanization, creates better living environments, brings about better services, develops the public's sense of responsibility, and sets new goals for cultural development. During this process, stakeholders will learn to exercise their rights, transforming from pure participants to both participants and organizers, and the market will see great opportunities for development. What the government should do is focusing more on amending related systems according to social changes occurring in built-environments, rather than spatial improvement or economic growth.

Aside from reconstruction or renovation which takes a long time, is there any other way that may create a better living environment for people living in old courtyard houses in hutong areas? This is a topic that Zang Feng, He Zhe, Shen Haien, and their People's Architecture Office have been studying since 2010. Small inner spaces, insufficient natural light, and no washing rooms are the biggest problems of courtyard houses. Through careful observation and analysis, People's Architecture Office designed a kind of low-cost, easy-to-assemble, energy efficient and comfortable boxes which have been placed in the houses of No. 37 Tiaozhou Hutong, and gradually the rest of Dazhalan.

These boxes are actually "plug-in" houses—a prefabricated modular construction system which provides a low-cost way to upgrade a house for a better living environment without pulling it down. An important part of the Dazhalan Renovation Program, "Plug-in" houses were designed to effectively protect and renovate Dazhalan, the nearest old block to Tiananmen. So far, the "plug-in" house has won a number of international awards, such as Architizer A+, WAF, Red Dot and Archmarathon.

Lying south of the Tiananmen Square, Dazhalan hasn't experienced large-scale demolition, and thus retains relatively compete narrow hutongs and old courtyard houses which, though being very rare in today's Beijing, undeniably have a number of problems, such as poor infrastructure, no washing rooms, lack of sewer pipes, insufficient natural light, and poor thermal insulation, sound insulation and damp proofing which cause inconvenience to the residents. How to protect these old houses while improving the residents' quality of life? To solve this problem, People's Architecture Office designed a plug-in prefabricated composite sheet modular construction model which integrates the structural system, thermal insulation system, pipe & cable systems, windows & doors, and interior & exterior decoration. After one year's experiments, the plug-in house system was included in the Dazhalan Renovation Program in October 2013, and 15 plug-in houses have been built in Dazhalan area so far.

Composite sheets are light which indicates a low transport cost. Besides, the assembly is very easy, requiring only some hex wrenches and several inexperienced people to build one plug-in house within one day. The plug-in house system can greatly improve the thermal insulation, sound insulation, and damp proofing of the original courtyard houses. What's more, the system only consumes 1/3 of the energy required for reconstruction, and

is only 1/2 and 1/5 of the costs for repair and reconstruction respectively. The system is equipped with diverse supplements,

城市更新从20世纪80年代由西方传入中国，最初的"新陈代谢""有机更新"强调城市物质环境的保护与发展。而21世纪以来，"城市再生""城市复兴"以及更多视角的理解则蕴含了文化、经济、社会、政治等多维度内涵。历史街区的城市更新牵涉多个利益主体，其顺利实施取决于多主体间的价值观和具体利益平衡。本演讲通过梳理广州城市更新实践的脉络，分析各个阶段特定与相应的决策机制，并结合相关案例来厘清政府、市场、居民等多个主体之间的权利责任与角色分工。在实证检讨的基础上，思辨城市更新未来的实践方向。

问题认知方面，目前历史街区的衰败可以从经济维度和社会维度来认识。从经济维度来看，城市中很多地区的区位价值释放后并没有被还原到建成环境中，呈现相当程度的利益外流，这导致建成环境因缺乏资金投入而走向衰败，要么被高强度地物质性置换，要么因低品质而不断物质性恶化。从社会维度来看，无论是拆除重建式物质性置换所带来的绅士化、商业化以及社会断裂，还是苟延残喘式物质性衰败所表现的贫困化和社会颓废，都"无可奈何"地存在着，并已经蔓延到20世纪八九十年代无力自维育的建成区。

在对广州城市更新历程回顾中可见，政府的角色在逐渐强化。政策演进的核心决策机制大致可以归纳为从市场驱动到政府包办，再到政府主导多元的三个阶段，中间出现过适度的纠错和探索，逐渐走向多元协同的理性过程。同时可见，政府在城市更新中逐渐从单一目标走向综合目标，并逐步将相关主体纳入到决策过程中。2016年在原"三旧"改造办公室基础上成立的广州城市更新局是国内首家正式的一级局，倡导以微改造微更新的方式实施，是一种具有城市复兴价值导向的城市更新，带有丰富的软性城市社会建设性质的内涵，而在这背后，又必须有完善的资金、责任的基本制度以及权利、决策的执行制度作为支撑。目前国内历史城区城市更新中还存在比较显著的问题是居民及公众参与仍处于初级阶段，显性物质性目标仍受到高度重视，而如何开展有效的参与，还有待实践积累经验。

我国新时期的城市更新，应是一个强调建成环境改善，同时不减损原有权利主体的权益，注重公平，倡导社会认同，并不断实现社区共同体自我更新和可持续发展的社会进步过程。对于政府，历史城区的城市更新需要更多关注建成环境的社会进程，以及相应的制度转型，而不仅仅是个空间改善或经济发展问题。对于市场，提供更高品质的生活环境和服务，以及更有责任感和城市文化目标的城市更新，也许是一个具有巨大潜力的新机遇。对于社会，快速城市化进程中，从积极参与到自组织地自下而上推进城市更新，则是一个主体培育、赋权到成熟的进程。

王世福
华南理工大学建筑学院教授、博士生导师，城市规划系主任
中国城市规划学会理事
富布赖特（Fulbright）麻省理工学院（MIT）高级访问学者

除了拆掉重建或者花很长时间改造，还有什么办法可以让胡同居民在老房子里住得更好？这是臧峰、何哲、沈海恩和他们的众建筑工作室2010年就开始研究的问题。他们观察老房子，思考并尝试解决空间狭小、采光不足、没有卫生间等难题。从他们工作室所在的苕帚胡同37号开始，一种低成本、易于安装、节能、环境舒适的"盒子"陆续被塞进北京大栅栏的胡同里。

内盒院本质就是"房中房"，是一个应用于旧城更新的预制化模块建造系统，提供一种既避免全拆重建、又相对低造价的方法，来提升人们的生活居住品质。内盒院是"大栅栏更新计划"的一个重要项目，旨在对这片距离天安门最近的历史街区进行有效的保护和更新。项目自推出以来屡获殊荣，包括Architizer A+、WAF、Red Dot及Archmarathon等国际奖项。

紧邻天安门广场南侧的大栅栏地区，没有经历过大规模的拆迁，仍保留有相对完整的狭窄胡同和老旧四合院，显得弥足珍贵。但同时，这里也长期存在着基础设施不完善，缺少卫生间下水管道，房屋密闭、保温、隔声、防潮不佳等问题，生活上有诸多不便。如何兼顾旧城保护与在地居民生活质量？众建筑开发出一种可内嵌于老旧房屋之中，集成了结构、保温、管线、门窗及室内外装饰完成面的预制复合板材；并在一年中将其由实验性的样板，完善成一套系统化的解决方案。自2013年10月携内盒院项目初次参加"大栅栏更新计划"至今，众建筑已在该地区内完成了15个内盒院项目。

板材质轻、运输便宜，安装也易操作，用一个六角扳手就可以把它们锁在一起。几个毫无专业技术训练的人，一天之内就能安装完成一个单独的内盒房子。经过如此改造的房屋，就具备了很好的保温与密闭性能，能耗约为新建四合院的1/3，造价约为修缮四合院的1/2、新建四合院的1/5。内盒院还有着多样化的选配插件，如夹层、伸缩屋、连通室内与院子的上翻屋、滑动墙、大平开墙

臧峰
众建筑/众产品联合创始人与主持建筑师之一
国家一级注册建筑师
北京大学建筑学硕士

such as the interlayer, telescopic room, upturning room which connects the original house and the yard, sliding wall, and large vertical hinged wall. There are two kinds of supplements to the washroom: the purification tank which turns sewage into reclaimed water, and the composting toilet. Residences may be attached with plug-in kitchens and washrooms, and offices may adopt a basic plug-in house system without supplements.

The plug-in house system mainly aims at scattered long-term unoccupied houses, and occupied houses needing upgrading without reconstruction. Residents may be subsidized to some extent for house renovation.

In China, many old blocks have been demolished directly. As residents were forced to leave, the original community relationships have been broken, and the hustle and bustle atmosphere has disappeared. Compared with such rude model driven by short-term interest, the plug-in house system provides a healthier renovation model pursuing long-term social interest. Residents may improve their quality of life by independently building energy efficient infrastructure without rebuilding their houses, or by relying on municipal infrastructure. Reconstruction needs lots of funds from a few investors, while the plug-in house system needs small amounts of investment from a large number of residents, which is more effective for promoting the long-term development of old blocks.

I'd like to say that it's my honor to participate in the design of Dazu Primary School Bud Class at Lugu Lake area, Yanyuan County, Sichuan Province, and Shuanglong Primary School at Xiuning County, Huangshan City. The two schools, which are all Hope Primary Schools adopting light steel frames, are similar in nature, function, structure and material, but their architectural expressions are quite different. Based on my experience and observation, I will analyze the architectural expressions and formation of the two designs to find the reason behind, and at the same time, discuss the relationship among the environment, process and expression of a design.

1 Dazu Primary School Bud Class at Lugu Lake, Yanyuan County, Sichuan Province

In a valley at Lugu Lake area where Sichuan and Yunnan share their borders, there lies Dazu Village which is home to both Naxi people and Mosuo people. Dazu Primary School is sitting among farmhouses built along the valley.

Buildings at Dazu Primary School adopt the timber frame where logs are laid horizontally and interlocked on the ends with notches. Consuming a large number of logs, these buildings also have poor daylight and thermal insulation. Unlike the old buildings, Bud Class adopts an innovative composite structure. Except C-shaped steel columns with vary small sections, all rod-shaped parts, which are thermal insulation broken bridge materials, can stabilize the structure. All doors and windows have been designed to be as wide as connected composite sheets so that they can be inserted in light steel frames together with the latter as a part of the whole structure. Parts forming the composite structure may be prefabricated, and assembled on site in a short period of time. Besides, the structure features good ventilation and thermal insulation which create a comfortable interior environment meeting modern requirements.

The design of Bud Class has also considered other factors such as terrain, climate, functions, space, and details, in a bid to incorporate the new building into the rest of the village in a harmonious way. Seen from the distance, the building, which is covered by red sheets, presents diverse changes in shape and more design details compared with other buildings in the village. Bud Class comprises three classrooms and one reading room. Square floor plans reduce not only the area for corridors, but also heat dissipation in winter. U-shaped glass walls separate neighboring rooms, increasing interior light while promoting the interactivity between different rooms. Arguably , extremely simple floor plans and proper materials can jointly create interesting interior spaces.

Widows, which look like photo frames directly connecting interior and exterior spaces, are structurally simple for easier manufacture, transport and construction. Air comes in from a small operable door below each window, and flows out from the operable door beside the roof light which allows more natural light into the room, forming good ventilation. In addition, the area between every two windows sinks to form a bookshelf, making the wall looks thicker which creates a sense of safety for those rural students. With advanced passive design, the building also adopts LED lighting and wind power generation technology.

2 Shuanglong Primary School at Xiuning County, Huangshan City

Located in Xiuning County, Huangshan City, Anhui Province, Shuanglong Primary School sits at the junction of Shuaishui River and Yangong River, with mountains stretching in the distance. As a trial project, the school has been designed to be multifunctional, for the purpose of maximizing the types of school activities, making full use of land and space, involving more local users, as well as promoting the communication between tourists and the locals, and attracting the attention and support from outside the village, thus building a multi-win relationship among all parties involved.

The project comprises a new building and a renovated building in the south and north of the project site respectively. The

等。卫生间插件也有两类：一类是将卫生间污水处理为中水的净化槽；另一类是无水堆肥马桶。如为居住空间，可以选择插入厨房和卫生间；如为办公空间，也可以选择没有插件的基本内盒空间。

内盒院主要针对已腾退但长期空置的零散房屋，以及希望提高居住质量但又不想重建房屋的本地居民。采用内盒院的居民有可能会得到一定的补贴，用于鼓励他们对自己房屋的修缮和投入。

在中国，很多老城区被不假思索地拆除，人们被迫离开家园，原有紧密的社区关联被切断，喧闹纷杂的历史图景被丢弃。相对于这种粗暴的短期利益驱动开发模式，内盒院则提供了一种追求长期社会利益的、更为健康的发展模式：居民们可以创建个人的、分散的、高效节能的基础设施，无需拆除房屋与依赖大市政基础设施即可直接提升居住质量。比起少数人的巨额投资，大量居民个人的微额投资反而会对这个地区的发展更为长期有效。

本人有幸参与了两个同以轻钢结构来完成的希望小学项目，分别是四川盐源泸沽湖达祖小学新芽学堂和黄山休宁双龙小学，两个项目在性质、功能以及结构材料都非常类同，而表达却有截然不同的效果。基于自己的参与和观察，我试图从设计过程中剖析两个设计表达的理念与形成，以说明其不同表达的由来，最后希望借此浅谈一下一个设计坏境、过程与最终表达的关系。

谭善隆
维思平建筑设计执行主设计师

1 四川达祖小学新芽学堂

四川与云南交界的泸沽湖区的峡谷中坐落着纳西族与摩梭族村寨——达祖村，山脚下的农舍间坐落着达祖小学。

达祖小学原有校舍采用木制井干结构，用原木咬合堆叠搭建而成，房屋的采光以及热工性能并不理想，也耗费大量的木材。而新芽学堂则采用了全新构想的复合结构体系。除了截面很小的 C 形钢柱外，所有被用来组成这个构造体系的杆件都是对结构稳定性起作用的，而且都做了断桥保温处理。窗、门也被设计成与复合材料相同的宽度，以便于同复合板一样插进轻钢框架中，并最后组合成结构整体。这个体系建造的建筑可以工厂预制、快速现场组装，室内通风、保温、隔热等方面的舒适性更容易达到现代要求。

新芽学堂同时根据具体的场地、气候、功能、空间、细节等方面的条件规划量体，尽力同达祖小学其他校舍及村庄融在一起。从近处看，建筑红色木板表面具有一种周围其他校舍及农舍所没有的层次及细节感。新芽学堂有三间教室和一个阅读室。方形的平面布局不但节约了走廊空间，也减少了冬季的散热量。房间之间 U 形玻璃隔墙使每间房间都可借到隔壁房间的光线，隔壁人的活动也隐约可见。一种极其简单的平面布局和有关的材料选择带给使用者内部空间的趣味。

窗户设计不但构造简单，便于制造、运输及建造，从室内看也十分通透，如同画框。建筑通风中的进风靠窗下一个可开启的小门解决，出风效果靠屋顶上采光天窗旁边的通风活门解决。这些天窗也将光线带到房间中部。窗洞四侧多出来的宽度伸进建筑内部，形成了窗户与窗户之间的书架，又带来建筑物墙壁厚重的错觉。这种"厚重感"满足了乡村小学生们寻求庇护的心灵需求。在良好的被动设计的基础上，新芽学堂还使用了 LED 照明和风力发电技术。

2 黄山休宁双龙小学

双龙小学位于安徽省休宁县，率水河与颜公河的交汇处，紧邻河岸，背靠群山。作为研究性的项目，设计师要的不止是一所学校，而是共享互动的希望小学：充分拓展小学的活动场地、发掘当地使用者，激发游客与当地交流，吸引外界机构的关注与支持，由学校带动村落，达到互动多赢。

小学包括基地南侧的新建建筑和北侧的改建建筑。新建建筑功能是位于建筑中部的 7 间教室和两端的活动空间，改建建筑承担生活和教学辅助功能。两栋建筑将场地划分为尺度不同的活动场地，既可容纳教学活动的大尺度场地，

former comprises seven class rooms in a row and two play areas at the ends. The later creates a space for recess and additional instruction. The two buildings divide the school into different areas varying in size, including bigger areas for teaching, and smaller areas for play. What's more, the school also allows public activities during after school hours.

The buildings have been designed and constructed in a very scientific way for future extensive copy. Most parts of the buildings were prefabricated in factories, and assembled on site. Recyclable building materials have been used to reduce energy consumption and pollution. The composite wall is formed by thermal insulation rock wool sandwich panels, light steel structure, gypsum boards, and water-proofing multiwall polycarbonate sheets (sunshine sheets) layer by layer from inside to outside. The space between insulation sheets and gypsum boards forms an air interlayer extending from the southern facade to the roof, promoting ventilation in summer and increasing thermal insulation in winter through opening/closing the vents. Multiwall polycarbonate sheets allow uniformly distributed diffused natural light into the rooms, avoiding shadows effectively. Shuanglong Primary School demonstrates well that an energy-efficient and comfortable interior environment can be created through rational design.

2.1 Similarities between the two projects

From customized design to mass production.
Project nature: charity, donation, Hope Primary School.
Project function: Primary schools, classrooms.
Main materials: Light steel frame profiles and other industrialized building materials.
Way of construction: Prefabrication + onsite assembly.
Project environment: Villages with difficult access.
Building size: One-story buildings adopting the column layout (1800 ~ 1900mm).

2.2 Differences between the two projects

Generally speaking, Dazu Primary School Bud Class doesn't express its industrialized manufacture through materials—it's modular design and standardized nodes silently create its regional architectural expressions. Shuanglong Primary School uses light steel and repeated structures to create a "folded" roof varying horizontally along the building. The former's expression focuses on spatial design and appearance, while the latter tries to express its structure and shape.

Dazu Primary School Bud Class was designed by a university which is a pure research environment, while Huanglong Primary School was designed by an architecture firm. Therefore, the former was carried out for the research of a specific topic, while the latter primarily aims at finding a solution for existing problems, with perfecting the design process as the minor objective.

Both of the two projects are trying to create an ideal architectural model for later extensive copy, in hope of providing a type of comfortable and sustainable buildings for users, and an efficient and reasonable way of construction for builders and developers.

3 Inspiration

The process of design is the process to solve a problem and express the design concept. The entry point will directly affect the solution and/or expression of a design. To find untypical solutions to a typical problem, the entry point may be creating different design environments and processes. In this way, although the solutions are similar substantially, the focuses of their expressions may vary greatly. This is one of the expirations that the two projects have given me.

Settlements are communities in which people live by solving life-related problems such as environmental protection, architecture, building technology, social construction, and interpersonal relationships using the power of nature, as well as their wisdom. How should we pass down such wisdom? What are the present and future pictures of our settlements? Those are topics to be discussed in Settlement.

又提供孩子自由活动的小尺度场地，在非教学时间，可提供村民举行公共活动。学校的设计与建造是以科学化且具推广性的设计与研究方法为基础。建筑大部分构件在工厂预制，现场组装搭建。建筑材料可循环利用，减少能耗及环境污染。外墙构造从室内到室外依次是：保温材料岩棉夹心板、轻钢结构、表皮及防水材料聚碳酸酯多层板（阳光板）。保温与表皮之间形成从南立面延伸到屋面的空气间层，通过开启和关闭通风口，夏季起到拔风作用；冬季达到保温作用。建筑采光均采用聚碳酸酯多层板，为教室提供均匀的漫射光，确保光线在投入教室后不会留下阴影。利用有效的设计手段，减少建筑能耗，创造舒适的室内环境。

2.1 两项目的类同处

从定制中思考量产。

项目性质：慈善，捐建，希望小学。

项目功能：小学，教室。

主要材料：轻型钢型材及其他工业化材料。

建造方式：异地生产构件，现场组装的施工方式。

基地环境：村落当中，交通相对不便的地区。

建筑尺度：一层，柱网约为1800～1900mm。

2.2 两项目的差异处

整体而言，达祖小学新芽学堂并不以它的材料来诉说它的工业化生产方式，它的模数化设计及标准节点是默默支撑着它地缘化的建筑表情。休宁双龙小学则把轻钢和其结构的重复规律来做基调，突出它"折纸"屋顶沿建筑长度的造型变化。前者表达止于空间组织和表皮，后者表达止于结构和造型。

两个设计在工作环境上也存在差异。达祖小学是在大学，即纯研究型的环境中完成的，而休宁双龙小学则是在建筑设计事务所中进行的。前者立案以研究题目主导项目，更可以理解是为研究题目来寻找项目；而事务所则是借项目来研究问题，以解答项目需求为主，以推敲设计方案的过程来进行研究。

两项目同样在思考一种设计的原型，可以说借定制设计来思考量产设计。希望给用户找到舒适、可持续的建筑形式，也给建造方、开发方找到高效又合理的构造方式。

3 启示

假如把一个设计理解成是对某问题的解题加上表达的话，我们如何去切入题目将直接改变我们的答案或表达，或同时影响两者。反过来看，要发现典型问题的非典型答案，我们也可以尝试透过创造不同的设计环境和过程来入手，尽管最终解题的内涵是类似的，其表达的重点和可能性也会大有变化，这是我从两所希望小学的设计中得到的启示之一。

聚落（settlement）是人类的聚居场所，是人类利用自然解决自己生活问题的结晶体。聚落的内部充满着人类如何解决环境问题，建筑问题，建造技术问题，社会问题以及解决其内部人与人之间关系问题的智慧。而这些智慧如何能够得以延续，聚落的现状与未来将如何发展和变迁，这一切将是"聚落"这个话题所要讨论的内容。

王昀
方体空间工作室主持建筑师

I am a fan of "6 + 1" project and make tiny contribution to this project during the "6 + 1" development process. However, I feel ashamed because my contribution is too tiny compared with the dedication, academic attitude and the public interest of the people from peer colleges. For this reason, I would like to take this opportunity to express my respect to the 6 +1 team.

The implementation of the graduation creation contest among universities with fixed proposition of design theme allows competition against each other, exchange of experience concerning various coaching style and concept from different universities across the country. Such contest is an admirable pattern that has a major impact on arousing the enthusiasm of teaching and learning, on the development of interior and environmental art teaching and scientific research, and on promotes friendship between universities.

"6 + 1" project, supported by CIID, is a pure land for academy and a hometown for dedication. The competition mechanism for this project has a stable assessment mechanism with a rigorous academic attitude. I wish and firmly believe that "6 + 1" will achieve more outcomes.

The students have attached much importance to the "local elements" and strictly complied with them with their compliance, which is conducive to the objectivity and relevance of the design process;

The students are sensitive to and meticulous in connection with "problem finding", which is quite important to the determination of "starting point" for the design;

"Proposal presentation" is deemed as the key step of the design event by the students, through of which they expect to gain more scores for the value of their overall design, reflecting more consensus has been made by and between the teachers and students in respect of a joint linking between education and career and of process-outcome unification, and the like.

The "position" for design shall be determined in a clearer and more resolute manner—we make design for others, not for self-interest;

The "purpose" of design shall be determined in a clearer and more resolute manner – we make design only for attainmentachievement of goal on an accurate and appropriate basis as the "better, faster and stronger" design makes no sense;

Practical and in-depth research shall be made upon the changing "Human behavior" and "formation of objects" – which is the drawback in terms of previous education and learning for design, and shall be focused on for today's education and learning.

我是一位"6+1"的粉丝。在"6+1"发展过程中从旁出过一点儿力,但是相较来自兄弟院校诸位先生的奉献精神、学术态度、公益心,我自愧弗如,借此机会,向"6+1"团队表示敬意。
　　校际之间开展命题毕业创作竞赛,使东南西北不同的执教风格、理念能够互相切磋、互相交流,这对于调动教与学的积极性,对于室内和环境艺术教学科研的发展,对于增进院校间友谊都是极好的模式。
　　由CIID支持的"6+1",其竞赛机制具有严谨的学术态度及相对稳定的评审机制,是一片学术净土和奉献的热土。我预祝也坚信"6+1"将会越办越好。

王炜民
中国美术学院教授

　　同学们对"在地要素"都有相当的重视和遵从,这对设计行进过程的客观性和针对性是有益的。
　　同学们对"发现问题"都表现得敏感和细致,这对设计"起点"的确定,是相当重要的。
　　同学们视"提案表述"为关键一环,都希冀借此环节为整个设计的价值加分,这体现了师生们在教育与职场的衔接、过程与结果的统一等方面,有了更多的共识。
　　要更清晰并坚决地解决设计的"站位"问题——为他人设计(而非按自我兴趣设计)。
　　要更清晰并坚决地解决设计的"控制"问题——设计不存在"更高更快更强"(而只能是"目标精准贴切")。
　　要更真切并深入地研究不断衍变着的"人的行为"与"物的业态"——这既是过往设计教与学的短板,也是当下设计教与学应着力的重点。

赵健
广州美术学院学术委员会主任,教授

歷史定格
Introduction of the Activity

CIID『室内设计 6+1』2017（第五届）
校企联合毕业设计
CIID"Interior Design 6+1"2017(Fifth Session)
University-Enterprise Cooperative Graduation Design

国匠承启卷
——传统民居保护性利用设计
Craftsmanship Heritage
—Design for the Protective Utilization of Traditional Folk Houses

柒

國匠承啓卷——傳統民居保護性利用設計

China Institute of Interior Design

Six colleges display differently and this reflects a certain difference between them.The predecessor of China Institute of Interior Design (CIID) is China Institute of Interior Architects. Since itsestablishment in 1989, CIID has been the only authorized academic institution in the field of interior design in China.CIID aims to unite interior architects of the whole country, raise the theoretical and practical level of China's interior design industry, pioneer the Chinese characteristics of interior design, help interior architects play their social role, preserve the rights and interests of interior architects, foster professional exchanges and cooperation with international peers, so as to serve and facilitate the construction of China's modernization.

Since its foundation 20 years ago, CIID hold abundant and colorful academic exchanges every year, building aplatform for designers to communicate and to study meanwhile update designer information of design industry, related competitions and business promotion, to enhance the better and rapid development of interior design industry of China.

Members of CIID are composed of individual members (including student members, associate members, fullmembers, advanced members, foreign members) and group members. By now, CIID has a large membership of more than ten thousands prominent designers who are from all over the country and passed the strict assessment by CIID.

Every year CIID will organize various types of competitions which include Institute Award of China Interior DesignAward, Influential People of China Interior Design, "Renewal Design" Original Competition, National-level Interior Design Competition for Young Students, China Hand-drawn Art Design Competition and so on.

CIID Secretarial is located in Beijing, taking charge of institute work. CIID secretariat publishes membership periodical china interior, Collection of Entries of China Interior Design, periodical Ornament and Decoration World,Home Adornments, ID+C. CIID website: www.ciid.com.cn.

Tongji University

Tongji University, established in 1907, is a top university of China Ministry of Education. During the time of restructuring of the university and college systems in 1952,the Department of Architecture was formed at Tongji University , and in 1986 was renamed as the College of Architecture Urban Planning (CAUP). Currently CAUP has three departments: the Department of Architecture, the Departmentof Urban Planning, the Department of Landscape Design. The undergraduate program covers:Architecture, Urban Planning, Landscape Design, Historic Building Protection and Interior Design. CAUP is one of China's most influential educational institutions with the most extensive programs among its peers, and the largest body of postgraduate students in the world. Today, CAUP hasbeen recognized as an international academic center with a global influence in the academic fields.

Tongji University's interior design education originated from the Department of Architecture which started to conduct interior space research in the 1950's.In 1959, it applied for the establishment of the "Interior Decoration and Furniture Specialty" within Architecture Discipline. In 1986,approved by the Ministry of Education and the Ministry of Construction, the "Interior Design Discipline" was formally founded. Starting to admit undergraduate studentsin 1987,Tongji University was one of two earliesthigh education institutions in mainland China to train interior design professionals in a University of science and technology. In 2011.

"Interior Design" officially became the secondary discipline of the Architecture Discipline. In the same year, the "Interior Design Research Team" was established,providing even broader room for subject development. Tongji University's interior design education crystallizes its own characteristics,emphasizing rational thinking and proposing the interior design concept of"human centric, ecological consciousness, overall environmental perspective,equal time and regional characteristic significance,technology and art integration".

CIID 中国建筑学会室内设计分会

中国建筑学会室内设计分会（CIID），前身是中国室内建筑师学会，成立于1989年，是在住房和城乡建设部中国建筑学会直接领导下、民政部注册登记的社团组织。CIID是获得国际室内设计组织认可的中国室内设计师的学术团体，是中国最具权威的室内设计学术组织。

学会的宗旨是团结全国室内设计师，提高中国室内设计的理论与实践水平，探索具有中国特色的室内设计道路，发挥室内设计师的社会作用，维护室内设计师的权益，发展与世界各国同行间的合作，为我国现代化建设服务。

CIID成立20多年来，每年举办丰富多彩的学术交流活动，为设计师提供交流和学习的场所，同时也为设计师提供丰富的设计信息，提供各类大型赛事信息，提供各项商务帮助，促进中国室内设计行业更好更快地发展。

CIID设有个人会员（包括学生会员、准会员、正式会员、资深会员、外籍会员）和团体会员。目前，已有会员1万余名，均是经过严格资格评审的精英设计师，遍布全国各地。CIID每年举办各类赛事，包括中国室内设计大奖赛"学会奖"，"中国室内设计影响力人物"评选，"设计再造"创意大赛，"新人杯"全国青年学生室内设计竞赛，中国手绘艺术设计大赛等一系列奖项。

CIID秘书处设在北京，负责学会相关工作。秘书处定期出版会员会刊《中国室内》以及《中国室内设计年刊》，同时学会拥有会刊《装饰装修天地》《家饰》《室内设计与装修 ID+C》。CIID的官方网站为中国室内设计网。

同济大学

同济大学创建于1907年，教育部直属重点大学。同济大学1952年在国家院系调整过程中成立建筑系，1986年发展为建筑与城市规划学院，下设建筑系、城市规划系和景观学系，专业设置涵盖城市规划、建筑设计、景观设计、历史建筑保护、室内设计等广泛领域。同济大学建筑与城市规划学院是中国大陆同类院校中专业设置齐全、本科生招生规模最大，世界上同类院校中研究生培养规模第一，具有全球性影响力的建筑规划设计教学和科研机构，是重要的国际学术中心之一。

同济大学室内设计教育起源于建筑系，同济大学建筑系于20世纪50年代就开始注重建筑内部空间的研究，1959年曾尝试在建筑学专业中申请设立"室内装饰与家具专门化"。1986年经国家建设部和教育部批准，同济大学建筑系成立了室内设计专业，1987年正式招生，成为中国大陆最早在工科类（综合类）高等院校中设立的室内设计专业。1996年原上海建材学院室内设计与装饰专业并入同济大学建筑系；2000年原上海铁道大学装饰艺术专业并入同济大学建筑系。2009年同济大学开始恢复建筑学专业（室内设计方向）的招生工作。2011年建筑学一级学科目录下，设立"室内设计"二级学科。

同济大学建筑城规学院的教学理念为以现代建筑的理性精神为灵魂，以自主创造、博采众长的学术品格为本色，以当代技术与地域文化的并重交融为导向，以国际学科前沿的跟踪交流为背景。室内设计教学突出建筑类院校室内设计教学特色，强调理性精神，提出"以人为本、关注生态、注重环境整体观、时代性和地域性并重、融科学性和艺术性于一体"的室内设计观。

South China University of Technology

South China University of Technology(SCUT), located in Guangzhou city, Guangdong province, was founded in 1934. It is a well-known Chinese university which has a long history and enjoys a high reputation. It is a national key university directly under the Ministry of Education of the People's Republic of China, one of the first national "211 project" and "985 project" key construction of colleges and universities.

The design institute of SCUT was founded in June 2010, with majors including industrial design, environmental design, information and interaction design, clothing and apparel design. The design institute closely relies on the strong advantage of technology and deep cultural heritage of SCUT and actively explores the way of highly industry integration and international cooperation, aiming to create famous heights for domestic and foreign design innovative talent training as well as design practice and service.

At present, the design institute grasps the development opportunity in design creativity industry, constantly renews education idea and makes bold exploration in design innovation personnel training mode. With "leading technology innovation, leading culture innovation, leading industry transformation and sustainable development" as the construction goal, a series of production and research platform including "creative and sustainable design and research institute" "space of contemporary art" "public platform of design experiment and practice" "interdisciplinary top creative talents cultivation test area" and "Tenglong research&development center" "cultural art and creative industry research center" "Chinese folk art research center" "ceramic culture research institute" has been built, striving to become the domestic leading design instutite with international influence, so as to support and lead the design industry development in guangdong and across the country.

Harbin Institute of Technology

Harbin Institute of Technology affiliates to the Ministry of Industry and Information Technology, and is among the first group of the national key universities to enter the national"211Project""985 Project" and to start the collaborative innovation "2011 plan". In order to train engineers, the Mid east railway authority founded the Harbin Sino Russian school in 1920, the predecessor of Harbin Institute of technology, which becomes the cradle of China's modern industry and technical personnel. The School has evolved into a distinctive, powerful, first class national key university, which is multidisciplinary, open, researchful and with international influence.

The discipline of Architecture in Harbin Institute of technology is one of the earliest architectural subjects in China, with more than 90 years'ups and downs. The school of Architecture has 4 undergraduate disciplines, including Architecture, Urban Planning, Landscape Architecture,Environmental Design, and 3 first-level disciplines, including Architecture, Urban and Rural Planning,Landscape Architecture, and secondary master's disciplines in Design. We have the first-level doctorate and master's authorization in Architecture, Urban and Rural Planning and Landscape Architecture, and secondary-discipline master's authorization in Design, and Post-doctoral Research Institute on architectural first-level discipline.With the cultural spirits of rigor and diligence,The school of Architecture has created a devoted, distinctive, qualified and dedicated teachers' team.We have gained distinctive and outstanding achievements in undergraduate teaching, postgraduate education and scientific research, and have formed our own academic characteristics in the Design of Public Buildings in Cold Region, Regional Architecture, Building Technology in Cold Region, Architectural History and Theory, Urban Planning and Designing in Cold Region and Environmental Design in Cold Region.

Xi'an University of Architecture and Technology

Located in the historical and cultural city Xi'an, covering an area of 4300 acres, Xi'an University of Architecture and Technology has beautiful campus environment and academic atmosphere. This university has quite a longhistory, which can be dated back to the Northern University, founded in 1895. Since then, in the higher education history of modern China, this university has been accumulating the first batch of disciplines essence in civil engineering, construction and environmental class. In 1956, this university was named as Xi'an Institute of Architectural Engineering. In 1959 and 1963, it was renamed as Xi'an Institute of Metallurgy and Xi'an Institute of Metallurgy

华南理工大学

华南理工大学位于广东省广州市，创建于1934年，是历史悠久、享有盛誉的中国著名高等学府。是中华人民共和国教育部直属的全国重点大学、首批国家"211工程""985工程"重点建设院校之一。

华南理工大学设计学院组建成立于2010年6月，现有工业设计、环境设计、信息与交互设计、服装与服饰设计等4个系。设计学院紧密依托华南理工大学雄厚的理工优势和深厚的人文底蕴，积极探寻与产业高度结合和国际化合作的道路，旨在打造享誉国内外设计创新人才培养和设计实践与服务的研究高地。

当前，设计学院紧紧把握设计创意产业的发展契机，不断创新 教育理念，大胆探索设计创新人才培养模式，以"技术创新引领、文化创意引领、产业转型引领、可持续发展引领"为建设目标，拥有"创意与可持续设计研究院"以及"当代艺术空间""设计实验与实践公共平台""跨学科拔尖创新人才培养试验区"和"腾龙研发中心""文化艺术与创意产业研究中心""中国民间艺术研究中心""陶瓷文化研究所"等一系列产学研平台，力争建设成为国内领先、有国际影响的设计学院，从而支撑、引领国家和广东设计产业发展。

哈尔滨工业大学

哈尔滨工业大学隶属于国家工业和信息化部，是首批进入国家"211工程""985工程"和首批启动协同创新"2011计划"建设的国家重点大学。1920年，中东铁路管理局为培养工程技术人员创办了哈尔滨中俄工业学校——即哈尔滨工业大学的前身，学校成为中国近代培养工业技术人才的摇篮。学校已经发展成为一所特色鲜明、实力雄厚，居于国内一流水平，在国际上有较大影响的多学科、开放式、研究型的国家重点大学。

哈尔滨工业大学建筑学学科是我国最早建立的建筑学科之一，历经90余载风雨砥砺。建筑学院建筑学现有建筑学、城市规划、景观学、环境设计4个本科专业和建筑学、城乡规划学、风景园林学3个一级学科博士、硕士学科硕士点。已获得建筑学、城乡规划学和风景园林学一级学科博士、硕士授予权，以及设计学二级学科硕士授予权，还设有建筑学一级学科博士后科研流动站。建筑学院始终秉持严谨治学、精于耕耘的文化精神，打造了一支朴实敬业、有特色、有能力、肯奉献的优秀教师团队。在本科教学、研究生培养及科学研究方面，特色鲜明，成绩显著。在寒地公共建筑设计、地域建筑设计、寒地建筑技术、建筑历史与理论、寒地城市规划与城市设计、寒地环境艺术设计等诸多方向上，均形成自己的学术特色。

西安建筑科技大学

西安建筑科技大学坐落在历史文化名城西安，学校总占地4300余亩，校园环境优美，办学氛围浓郁。学校办学历史源远流长，其办学历史最早可追溯到始建于1895年的北洋大学，积淀了我国近代高等教育史上最早的一批土木、建筑、环境类学科精华。1956年，时名西安建筑工程学院。1959年和1963年，曾先后易名为西安冶金学院、西安冶金建筑学院。1994年3月8日，经国家教委批准，更名为西安建筑科技大学，是公认的中国最具影响力的土木建筑类院校之一及原冶金部重点大学。

and Construction. On March 8, 1994, approved by the State Board of Education, it was renamed as Xi'an University of Architecture and Technology and was recognized as one of China's most influential civil engineering colleges and the key university of the former Ministry of colleges and the key university of the former Ministry of Metallurgical.

Featured by civil engineering, construction, environment and materials science, engineering disciplines as the main body, Xi'an University of Architecture and Technology is a multidisciplinary university also with liberal arts, science, economics, management, arts, law and other disciplines. The university has 16 departments, 60 undergraduate programs so it can launch the first batch of undergraduate enrollment. It also has the right to recruit students by recommendation and the right of implementation of Accelerated Degree. Undergraduate art and design program is the featured major in Shaanxi Province.

Founded in April, 2004, Xi'an University of Architecture and Technology was established by the undergraduates from the major of art design and photography and from mechanical and electrical engineering industrial design and the relevant teachers from newly established sculpture and other specialties. The current undergraduate majors in this college include art and design, industrial design, photography, sculpture, exhibition art and technology, with more than 1,200 undergraduate students. Art Design was named "national characteristic specialty", "provincial famous professional". This university has gathered many multidisciplinary researchers, including architecture, planning, landscape, etc. All these research teams have a long history of working towards the research of western region cultures, through undertaking many national and provincial funds subjects. The Arts College has actively organized (or as the contractor) the national academic, discipline-building meetings; inviting international and domestic famous professors to come for academic exchanges. It also has developed management approach, and set up a special fund to encourage young teachers and outstanding doctoral students to carry out academic exchanges and international (inside) collaborative researches. In the meantime it has established friendly and cooperative relations with the universities in Europe, Asia and other countries.

The university has taken the overall development of students as its training objectives, the improvement of theoverall quality of them as the aim to focus on. Relying on various student organizations carrier and platforms, the university has carried out various forms of extracurricular activities. And also it has focused on strengthening academic exchanges and interaction, inviting scholars, experts and celebrities to come to listen to the lectures and presentations, which can broaden the students' horizons, improve their knowledge structure and culture their spirits of science, technology and humanities. In other ways, the university organized the students to actively participate in academiccompetitions, and guided or encouraged students to engage in research activities, and many students have published various papers in the national magazines. The college has transferred departments, libraries, laboratories and paid multi-interactive efforts or work together to build a teaching-research-student trinity open experiment (work) platform. The graduates trained by the college have been welcomed by employers and the graduates are in short supply.

Beijing University of Civil Engineering and Architecture

Beijing University of Civil Engineering and Architecture (BUCEA) is a university jointly built by the People's Government of Beijing Municipality and the Ministry of Housing and Urban-Rural Development, a pilot university for the "Excellent Engineer Education and Training Program" of the Ministry of Education and an advanced university of Beijing for the Party building and ideological and political work. It is a multiversity with distinct architectural characteristics and centering on engineering courses, the base of Beijing for talent cultivation and technology service in urban planning, construction and management, the base of Beijing for climate change study and talent cultivation, the national base for architectural heritage protection study and talent cultivation, and the only architecture institution of higher education in Beijing.

The predecessor of BUCEA was Beijing Industrial Engineering School established by the Qing Government in 1907. In April 2013, it was renamed BUCEA upon the approval of the Ministry of Education. BUCEA recovered undergraduate recruitment in 1977; was identified as one of the bachelor's degree awarding universities of the first batch in China in 1982; and was identified as a pilot university for the "Excellent Engineer Education and Training Program" of the Ministry of Education in 2011. In 2012, BUCEA's "Architectural Heritage Protection Theories and Techniques" was approved as a doctoral talent cultivation program to serve the State's special needs and BUCEA became a doctoral talent cultivation unit. In 2014, BUCEA obtained the approval to establish the Architecture Center for Post-doctoral Studies. In October 2015, the People's Government of Beijing Municipality and the Ministry of Housing and Urban-Rural Development signed an agreement for jointly developing BUCEA, and BUCEA was thus formally included in the list of universities co-sponsored by province and ministry. In May 2016, BUCEA's Sophisticated Innovation Center for Future Urban Design was approved as the Sophisticated Innovation Center of Beijing Institutions of Higher Education.

BUCEA has two campuses, one in Xicheng District and one in Daxing District. At present, BUCEA is accelerating construction of these two campuses in accordance with the development goal of "developing Daxing campus into a high-quality

西安建筑科技大学是以土木、建筑、环境、材料学科为特色，工程学科为主体，兼有文、理、经、管、艺、法等学科的多科性大学。学校现有16个院（系），其60个本科专业面向全国第一批招生，有权招收保送生，实行本硕连读。艺术设计本科专业为陕西省特色专业。

西安建筑科技大学艺术学院成立于2002年4月，是由建筑学院的艺术设计专业和摄影专业本科生、机电工程学院工业设计专业本科生和新成立的雕塑专业及各专业关教师组建而成。学院现有艺术设计、工业设计、摄影、雕塑、会展艺术与技术5个本科专业，在校本科生1200余人。艺术设计专业被评为"国家级特色专业""省级名牌专业"。学院集聚了包括建筑、规划、景观等在内的多学科的研究人才，学科团队长期致力于西部地区地域文化研究，承担了多项国家、省部级基金课题。艺术学院积极主办（承办）国家级学术、学科建设会议；邀请国际、国内知名教授来我校进行学术交流；制定管理办法，并设立专项基金，鼓励青年教师和优秀博士生开展学术交流、国际（内）合作研究，与欧洲、亚洲地区的多所大学建立了友好合作关系。

学院以学生全面发展为培养目标，注重学生综合素质提高，依托各类学生组织载体和平台，开展形式多样的课外活动。注重加强学术交流与互动，邀请学者、专家和社知名人士来我院举办讲座和专题报告，开阔学生视野，改善学生知识结构，培养学生的科技、人文精神。组织学生积极参与学科竞赛，指导、鼓励学生从事科研活动，在国内刊物上发表各类论文。学院调动教研室、资料室、实验室，多方互动，通力合作，构建了教学、科研、学生三位一体的开放性实验（工作）平台。学院培养的学生深受用人单位欢迎，毕业生供不应求。

北京建筑大学

北京建筑大学是北京市和住房城乡建设部共建高校、教育部"卓越工程师教育培养计划"试点高校和北京市党的建设和思想政治工作先进高校，是一所具有鲜明建筑特色、以工为主的多科性大学，是"北京城市规划、建设、管理的人才培养基地和科技服务基地"、"北京应对气候变化研究和人才培养基地"和"国家建筑遗产保护研究和人才培养基地"，是北京地区唯一一所建筑类高等学校。

学校肇始于1907年清政府成立的京师初等工业学堂。2013年4月经教育部批准更名为北京建筑大学。学校1977年恢复本科招生，1982年被确定为国家首批学士学位授予高校，1986年获准为硕士学位授予单位。2011年被确定为教育部"卓越工程师教育培养计划"试点高校。2012年"建筑遗产保护理论与技术"获批服务国家特殊需求博士人才培养项目，成为博士人才培养单位。2014年获批设立"建筑学"博士后科研流动站。2015年10月北京市人民政府和住房城乡建设部签署共建协议，学校正式进入省部共建高校行列。2016年5月，学校"未来城市设计高精尖创新中心"获批"北京高等学校高精尖创新中心"。

学校有西城和大兴两个校区。目前，学校正按照"大兴校区建成高质量本科人才培养基地，西城校区建成高水平人才培养和科技创新成果转化协同创新基地"的"两高"发展布局目标加快推进两校区建设。与住建部共建中国建筑图书馆，是全国建筑类图书种类最为齐全的高校。

学校坚持立德树人，培育精英良才。现有各类在校生11645人，已形成

undergraduate talent cultivation base and Xicheng campus into a collaborative innovation base for cultivation of high-level talents and transformation of science & technology innovation achievements. Additionally, BUCEA has joined hands with the Ministry of Housing and Urban-Rural Development to build China Architecture Library, and now it is the university that owns most complete categories of architectural books nationwide.

BUCEA has been adhering to the principle of "strengthen moral education and cultivate elites". Currently, BUCEA has 11,645 current students and has formed an all-round and multi-level school layout and education system covering undergraduate education, postgraduate education, PhD education, post-doctor education, full-time education, adult education and international student education. Over the years, BUCEA has cultivated more than 60,000 excellent graduates for the country, who have participated in Beijing's key urban construction projects in the past 60 years and have become the backbone of the national and Beijing's urban construction systems. BUCEA's graduate employment rate has been maintaining at above 95% over the years, and it was included in the list of "China's Top 50 Universities for Graduate Employment" in 2014.

BUCEA, facing the international market, features diverse education approaches. It has been insisting on the strategy of "running school openly" to widely carry out international exchange and cooperation in education. So far, it has established interschool exchange and cooperation relations with 60 universities of 26 countries and regions including the USA, France, the UK, and Germany.

Standing at a new historical starting point and centering on the basic task of "strengthen moral education", BUCEA is comprehensively promoting connotation construction, deepening integral reform, implementing managing university by law and reinforcing Party building in accordance with the working policy of "quality improvement, transformation, and upgrade". It has been enhancing its school-running strength, core competitiveness and social influence to drive its every undertaking to a higher level with the standard of "to be the best", and it is advancing towards the ambitious goal of "developing BUCEA into a domestic first-class, internationally well-known, high-level, open and innovative university with distinct architectural features".

Nanjing University of the Arts

Nanjing University of the Arts is one of the earliest arts institutions in China. It consists of 14 schools, 27 undergraduate majors and 50 major directions. It has Master's and Doctoral degrees and Post-doctoral stations in 5 subdisciplines under the first-class discipline of thearts: Arts Theory, Music and Dance, Drama and Film, Fine Arts Theory and Design Theory.

The professional background: in 2005, Display Design was set up as a major direction in undergraduate level and a research direction in master program in Nanjing University of the Arts; in 2008, it was incorporated into industrial design major as its one direction in School of Industrial Design; in 2011, it was approved by the Ministry of Education as an independent sub-discipline in then national disciplinary classification in undergraduate education; in 2012, it was classified into the first-class discipline of design with a new major name of "Art and Technology ".Through nearly 10 years of efforts by adhering to the principle that is students centered, academy-oriented, practice focused and development-guided, Art and Technology (Display Design) major has formed a coherent and open modular curriculum system of coherent knowledge and rational structure supported by modernization and globalization oriented course contents. The major is to cultivate professional design talents for the cultural sector,the museum sector, medium and large exhibition halls, the design community, the tourism sector, exhibition and other institutions. The graduates of this major are to have the ability to do design and research in the manner of integrating question, market and culture. And they are also to be cultivated as talents with the capacity of high-level artistic formation and excellent expression, as well as rational expertise structure and outstanding professional features.

从本科生、硕士生到博士生和博士后，从全日制到成人教育、留学生教育全方位、多层次的办学格局和教育体系。多年来，学校为国家培养了 6 万多名优秀毕业生，他们参与了北京 60 年来重大城市建设工程，成为国家和首都城市建设系统的骨干力量。学校毕业生全员就业率多年来一直保持在 95% 以上，2014 年进入"全国高校就业 50 强"行列。

学校面向国际，办学形式多样。学校始终坚持开放办学战略，广泛开展国际教育交流与合作。目前已与美国、法国、英国、德国等 26 个国家和地区的 60 所大学建立了校际交流与合作关系。

站在新的历史起点上，学校正按照"提质、转型、升级"的工作方针，围绕立德树人的根本任务，全面推进内涵建设，全面深化综合改革，全面实施依法治校，全面加强党的建设，持续增强学校的办学实力、核心竞争力和社会影响力，以首善标准推动学校各项事业上层次、上水平，向着把学校建设成为国内一流、国际知名、具有鲜明建筑特色的高水平、开放式、创新型大学的宏伟目标奋进。

南京艺术学院

南京艺术学院是我国独立建制创办最早并延续至今的高等艺术学府。下设 14 个二级学院，27 个本科专业及 50 个专业方向。拥有艺术学学科门类下设的艺术学理论、音乐与舞蹈学、戏剧与影视学、美术学以及设计学全部 5 个一级学科的博士、硕士学位授予权及博士后科研流动站。

南京艺术学院从 2005 年开设了展示设计本科专业和硕士专业研究方向；2008 年该专业并入工业设计学院，2011 年会展艺术与技术专业作为独立的二级学科获得国家教育部的正式批准，2012 年该专业又被归为设计学类，成为"艺术与科技"专业。南京艺术学院工业设计学院的艺术与科技（展示设计）专业以学生为中心，以学术为导向，以实践为手段，以发展为目标，通过近 10 年的发展，已经逐步形成知识融贯、结构合理、连贯而开放的模块化专业课程体系和走向现代化、全球化的课程内容。旨在为文化部门、博物馆部门、大中型展馆、设计团体、旅游部门、会展机构等单位培养具有一定的理论素养，专业知识合理，专业特点突出，具备问题导入、市场导入和文化导入的整合设计和研究能力，以及高度艺术造型及表达能力的专业设计人才。

Zhejiang University of Technology

Zhejiang University of technology is a key comprehensive college of the Zhejiang Province; its predecessor can be traced back to the founding in 1910 as Zhejiang secondary industrial school.After several generations' hard working and unremitting efforts, the school now has grown to be a comprehensive University in teaching and researching which is very influential. The comprehensive strength ranks the top colleges and universities. In 2009, Zhejiang province people's government and the Ministry of education signed a joint agreement; Zhejiang University of Technology became the province ministry co construction universities.

In 2013 Zhejiang University of Technology led the construction of Yangtze River Delta green pharmaceutical Collaborative Innovation Center which was selected for the national 2011 program,to become one of the first 14 of 2011 collaborative innovation center. There are 68 undergraduate schools; 101 grade-2 subjects of master's degree authorization; 25 grade-2 subjects of doctor's degree authorization; 4 postdoctoral research stations. Subjects include philosophy, economics, law,education, literature, science, engineering, agriculture, medicine, management, arts and other 11 categories. School teacher is strong. There are 2 Chinese academicians of Academy of Engineering,sharing 3 academicians of Chinese Academy of Sciences and Academy of Engineering; 6 national young experts with outstanding contributions; 3 National Teaching Masters, 3 winners of national outstanding youth fund, 2 people were selected to central thousand person plan, the Ministry of education,1 professor of the Yangtze River scholars, 1 innovative team of Ministry of Education, 2 national teaching teams, and 26 person were selected to all kinds of national personnel training plans.Zhejiang University of Technology adheres to its motto "Profound accomplishment and invigorating practice. Accumulate virtues and good practice." To improve the quality of education in a prominent position, and strive to cultivate to lead, promote Zhejiang and even the country's economic and social development of elite talent.

浙江工业大学

浙江工业大学是一所教育部和浙江省共建的省属重点大学，其前身可以追溯到1910年创立的浙江中等工业学堂。经过几代工大人的艰苦创业和不懈奋斗，学校目前已发展成为国内有一定影响力的综合性的教学研究型大学，综合实力稳居全国高校百强行列。

2013年浙江工业大学牵头建设的长三角绿色制药协同创新中心入选国家2011计划，成为全国首批14家拥有"2011协同创新中心"之一的高校。目前学校有本科专业68个；硕士学位授权二级学科101个；博士学位授权二级学科25个；博士学位授权一级学科5个；博士后流动站4个。学科涵盖哲学、经济学、法学、教育学、文学、理学、工学、农学、医学、管理学、艺术学等11大门类。学校师资力量雄厚，拥有中国工程院院士2人、共享中国科学院和中国工程院院士3人、国家级有突出贡献中青年专家6人、国家级教学名师3人、国家杰出青年基金获得者3人、中央千人计划入选者2人、教育部长江学者特聘教授1人、教育部创新团队1个、国家级教学团队2个、各类国家级人才培养计划入选者26人次。浙江工业大学坚持厚德健行的校训，把提高教育质量放在突出位置，努力培养能够引领、推动浙江乃至全国经济和社会发展的精英人才。

致 谢
Acknowledgements

中国建筑学会室内设计分会第九（广州）专业委员会
中国建筑学会室内设计分会第七（杭州）专业委员会
广州集美组设计机构

集艾室内设计（上海）有限公司
AAUA 亚洲城市与建筑联盟
哈尔滨工业大学城市规划设计研究院规划院设计五所
荣禾集团德和建筑设计事务所有限公司
北京歌华设计有限公司
南京市民俗博物馆
杭州国美建筑设计研究院有限公司

华南理工大学建筑学院教授、博士生导师，中国城市规划学会理事 王世福
谢英俊建筑师事务所主持设计师 谢英俊
深圳市公共艺术中心主任 黄伟文
众建筑 / 众产品创始合伙人主持设计师 臧峰
都市意匠城镇规划设计（北京）中心总建筑师 宋刚
维思平建筑设计执行主设计师 谭善隆
方体空间工作室主持建筑师 王昀

中国建筑学会室内设计分会资深顾问、广州美术学院原副院长 赵健
杭州国美建筑设计研究院有限公司董事长 王炜民
北京筑邦建筑装饰工程有限公司设计中心总经理 张磊
荣禾集团德和建筑设计事务所有限公司总经理设计总监 寇建超
AAUA 亚洲城市与建筑联盟秘书长 姚领